Eleventh International
Congress of Anatomy, Part C

BIOLOGICAL RHYTHMS IN STRUCTURE AND FUNCTION

PROGRESS IN CLINICAL AND BIOLOGICAL RESEARCH

Series Editors
Nathan Back
George J. Brewer

Vincent P. Eijsvoogel
Robert Grover
Kurt Hirschhorn

Seymour S. Kety
Sidney Udenfriend
Jonathan W. Uhr

RECENT TITLES

Vol 50: **Rights and Responsibilities in Modern Medicine: The Second Volume in a Series on Ethics, Humanism, and Medicine,** Marc D. Basson, *Editor*

Vol 51: **The Function of Red Blood Cells: Erythrocyte Pathobiology,** Donald F. H. Wallach, *Editor*

Vol 52: **Conduction Velocity Distributions: A Population Approach to Electrophysiology of Nerve,** Leslie J. Dorfman, Kenneth L. Cummins, and Larry J. Leifer, *Editors*

Vol 53: **Cancer Among Black Populations,** Curtis Mettlin and Gerald P. Murphy, *Editors*

Vol 54: **Connective Tissue Research: Chemistry, Biology, and Physiology,** Zdenek Deyl and Milan Adam, *Editors*

Vol 55: **The Red Cell: Fifth Ann Arbor Conference,** George J. Brewer, *Editor*

Vol 56: **Erythrocyte Membranes 2: Recent Clinical and Experimental Advances,** Walter C. Kruckeberg, John W. Eaton, and George J. Brewer, *Editors*

Vol 57: **Progress in Cancer Control,** Curtis Mettlin and Gerald P. Murphy, *Editors*

Vol 58: **The Lymphocyte,** Kenneth W. Sell and William V. Miller, *Editors*

Vol 59: **Eleventh International Congress of Anatomy,** Enrique Acosta Vidrio, *Editor-in-Chief*
Published in 3 Volumes:
  Part A: **Glial and Neuronal Cell Biology,** Sergey Fedoroff, *Editor*
  Part B: **Advances in the Morphology of Cells and Tissues,** Miguel A. Galina, *Editor*
  Part C: **Biological Rhythms in Structure and Function,** Heinz von Mayersbach, Lawrence E. Scheving, and John E. Pauly, *Editors*

Vol 60: **Advances in Hemoglobin Analysis,** Samir M. Hanash and George J. Brewer, *Editors*

Vol 61: **Nutrition and Child Health: Perspectives for the 1980s,** Reginald C. Tsang and Buford Lee Nichols, Jr., *Editors*

See pages 243–244 for previous titles in the series.

Eleventh International Congress of Anatomy, Part C
BIOLOGICAL RHYTHMS IN STRUCTURE AND FUNCTION

Proceedings Sponsored by the International Federation of
Associations of Anatomists and the Mexican Society of Anatomy
August 17-23, 1980, Mexico City, Mexico

Editor-in-Chief
ENRIQUE ACOSTA VIDRIO
President of the National
Organizing Committee
President of the International
Federation of Anatomy

Editors
HEINZ VON MAYERSBACH
Department of Anatomy
Medizinische Hochschule Hannover
Hannover, West Germany

LAWRENCE E. SCHEVING
Department of Anatomy, College of Medicine
University of Arkansas for Medical Sciences
Little Rock, Arkansas

JOHN E. PAULY
Department of Anatomy, College of Medicine
University of Arkansas for Medical Sciences
Little Rock, Arkansas

ALAN R. LISS, INC. • NEW YORK

Address all Inquiries to the Publisher
Alan R. Liss, Inc., 150 Fifth Avenue, New York, NY 10011

Copyright © 1981 Alan R. Liss, Inc.

Printed in the United States of America.

Under the conditions stated below the owner of copyright for this book hereby grants permission to users to make photocopy reproductions of any part or all of its contents for personal or internal organizational use, or for personal or internal use of specific clients. This consent is given on the condition that the copier pay the stated per-copy fee through the Copyright Clearance Center, Inc., 21 Congress Street, Salem, MA 01970, as listed in the most current issue of "Permissions to Photocopy" (Publisher's Fee List, distributed by CCC, Inc.) for copying beyond that permitted by sections 107 or 108 of the US Copyright Law. This consent does not extend to other kinds of copying, such as copying for general distribution, for advertising or promotional purposes, for creating new collective works, or for resale.

Library of Congress Cataloging in Publication Data
International Congress of Anatomy (11th: 1980:
   Mexico City, Mexico)
   Eleventh International Congress of Anatomy: proceedings.
   (Progress in clinical and biological research; 59)
   Includes index.
   Contents: pt. A. Glial and neuronal cell biology—pt. B. Advances in the morphology of cells and tissues—pt. C. Biological rhythms in structure and function.
   1. Neurobiology—Congresses. 2. Neuroanatomy—Congresses. I. Acosta Vidrio, Enrique. II. International Federation of Associations of Anatomists. III. Mexican Society of Anatomy. IV. Series.
[DNLM: 1. Neuroglia—Congresses. 2. Neurology—Congresses. W1 PR668E v. 59A / WL 102 G558 1980]
QP351.I76 1980   599.01'88   81-2778
ISBN 0-8451-0059-9 (set)   AACR2

Prof. Dr. med. Heinz von Mayersbach
1921–1980

Dedication

Heinz von Mayersbach was born an Austrian citizen in Meran, Südtirol, Italy, on February 13, 1921. He started his scientific career as an assistant in the Institut für Histologie und Embryologie in Graz in 1945, and in 1951 he received an M.D. degree from the University of Graz. He later served as British Council Bursar in the Division of Histochemistry of the Department of Pathology at the Royal Postgraduate School in London (1955–1956); Universitätsdozent at the Institut für Histologie und Embryologie in Graz (1957–1960); Chef de Travaux at the Institute for Histology and Embryology at the University of Lausanne (1960–1962); Professor and Director of the Institutes for Cytology and Histology of the University of Nijmegen in The Netherlands (1962–1966); and Professor and Director of the Department of Anatomy at the Medizinische Hochschule Hannover from 1966 until his death in 1980.

Professor von Mayersbach was known for his work in theoretical and applied histochemistry and immunohistochemistry, protein transport between mother and fetus, and, from 1960 on, chronobiology. In the last field he focused his attention in circannual and seasonal variations in liver morphology and physiology and on the effect of drugs applied at different circadian phases.

Professor von Mayersbach was author, coauthor, or editor of six books, and served on the editorial boards of five journals. His accomplishments over the years brought him honors and recognition, including the prestigious Mitgleid der Deutschen Akademie der Naturforscher "Leopoldina." He was appointed Chairman of its Section on Anatomy in 1979. He received a Commenius Medaille in 1966, and was President of the Gesellschaft für Histochemie the same year. He **was made a fellow of the Royal Microscopical Society (Oxford) in 1967, an Honorary Member of the Histochemischen Gesellschaft in 1970, and Honorary President of the International Society for Chronobiology in 1979.**

Professor Dr. med. Heinz von Mayersbach was indeed "einen warmherzigen Menschen, einen grossen Wissenschaftler und eine überragende Persönlichkeit." This volume of contributions to the subject he loved so much is affectionately dedicated to his memory.

Contents

Contributors ix

Contents of Eleventh International Congress of Anatomy, Parts A and B xi

Preface
Lawrence E. Scheving and John E. Pauly xv

An Introduction to Chronobiology
John E. Pauly 1

RHYTHMS IN CELLS AND TISSUES

Chronomorphology of Mammalian Tissues
H.v. Mayersbach and K.M.H. Philippens 25

Circadian Rhythms in Cell Proliferation: Their Importance When Investigating the Basic Mechanism of Normal Versus Abnormal Growth
Lawrence E. Scheving 39

Circadian Variation in the Proportion of Cells in Cell Cycle Phases in Hamster Tongue Epithelium Measured by Flow Microfluorometry
Norma H. Rubin 81

Chronobiological Aspects of Bone Marrow and Blood Cells
Ole Didrik Laerum and Nils Petter Aardal 87

Circadian Rhythmic Variations of the Relative Number of Binucleated Liver Cells in Rats
K.M.H. Philippens, S. Röver, and J. Abicht 99

Fluctuations in Nuclear and Cytoplasmic Size of Vaginal and Buccal Epithelial Cells Reflect the Time of the Ovulation as Well as the Time of the Day
W.J. Rietveld and M.E. Boon 109

CHRONOHISTOCHEMISTRY

Circadian Changes of Lysosomal Enzyme Activities in Rat Hepatocytes Using Ultracytochemistry
Yasuo Uchiyama and Heinz von Mayersbach 117

Lysosomal Histochemistry in Relation to a Synchronizer
Radium Dalwadi Bhattacharya 125

RHYTHMS IN SUSCEPTIBILITY

Circadian System and Teratogenicity of Cytostatic Drugs
 Ingrid Sauerbier 143
Circadian Stage Dependence in Radiation: Response of Dividing Cells *In Vivo*
 Norma H. Rubin 151
Circadian Host and Tumor Rhythms in Balb/C Mice. Rhythm Induction in Harding-Passey Melanoma
 L.L. Sackett, E. Haus, D. Lakatua, and J. Swoyer 165

CIRCANNUAL RHYTHMS

Nyctohemeral and Seasonal Variations in the Number of Tritiated Thymidine Labelled Cells in the Epiphyseal Cartilage of the Tibia in the Growing Rat. Effect of Lighting Duration and Temperature
 C. Oudet and A. Petrovic 187
Seasonal Variations in the Direction of Growth of the Mandibular Condyle
 A. Petrovic, J. Stutzmann, and C. Oudet 195

CONTROL OF RHYTHMS

The Role of Suprachiasmatic Nucleus Afferents in the Central Regulation of Circadian Rhythms
 W.J. Rietveld and G.A. Groos 205
Timing of the Estrous Cycle in Rats. Endogenous Peroxidase Activity in the Hypothalamic Arcuate Nucleus as a Tool in Circuit Analysis
 W.J. Rietveld, E. Marani, and J.C. Osselton 213
Chronobiological Aspects of the Mammalian Pineal Gland
 Russel J. Reiter 223

Index 235

Contributors

Nils Petter Aardal, Department of Pathology, The Gade Institute, University of Bergen, 5016 Haukeland Hospital, Norway **[87]**
J. Abicht, Department of Anatomy, Medizinische Hochschule Hannover, Hannover, Federal Republic of Germany **[99]**
Enrique Acosta Vidrio, President of the National Organizing Committee of the Eleventh International Congress of Anatomy and President of the International Federation of Anatomy
Radium Dalwadi Bhattacharya, Department of Physiology, B.J. Medical College, Ahmedabad 380 016, India **[125]**
M.E. Boon, Department of Pathology, SSDZ, Delft, The Netherlands **[109]**
G.A. Groos, Department of Physiology and Physiological Physics, University of Leiden, Wassenaarseweg 62, 2300 RC Leiden, The Netherlands **[205]**
E. Haus, Department of Anatomic and Clinical Pathology, St. Paul-Ramsey Medical Center, St. Paul, Minnesota 55101 **[165]**
Ole Didrik Laerum, Department of Pathology, The Gade Institute, University of Bergen, 5016 Haukeland Hospital, Norway **[87]**
D. Lakatua, Department of Anatomic and Clinical Pathology, St. Paul-Ramsey Medical Center, St. Paul, Minnesota 55101 **[165]**
E. Marani, Department of Physiology and Physiological Physics, University of Leiden, Wassenaarseweg 62, 2300 RC Leiden, The Netherlands **[213]**
Heinz von Mayersbach, Department of Anatomy, Medizinische Hochschule, Hannover, Federal Republic of Germany **[25, 117]**
J.C. Osselton, Department of Anatomy, University of Leiden, Wassenaarseweg 62, 2300 RC Leiden, The Netherlands **[213]**
C. Oudet, F.R.A. 15, INSERM, Institut de Physiologie, Faculté de Médecine, 4 rue Kirschleger, 67085 Strasbourg Cedex, France **[187, 195]**
John E. Pauly, Department of Anatomy, College of Medicine, University of Arkansas for Medical Sciences, Little Rock, Arkansas 72205 **[xv, 1]**
A. Petrovic, F.R.A. 15, INSERM, Institut de Physiologie, Faculté de Médecine, 4 rue Kirschleger, 67085 Strasbourg Cedex, France **[187, 195]**

The number in bold type following the contributor's affiliation is the first page number of that contributor's article.

K.M.H. Philippens, Department of Anatomy, Medizinische Hochschule Hannover, Hannover, Federal Republic of Germany [25, 99]

Russel J. Reiter, Department of Anatomy, The University of Texas Health Science Center, 7703 Floyd Curl Drive, San Antonio, Texas 78283 [223]

W.J. Rietveld, Department of Physiology and Physiological Physics, University of Leiden, Wassenaarseweg 62, 2300 RC Leiden, The Netherlands [109, 205, 213]

S. Röver, Department of Anatomy, Medizinische Hochschule Hannover, Hannover, Federal Republic of Germany [99]

Norma H. Rubin, Department of Human Biological Chemistry and Genetics, Division of Cell Biology, University of Texas Medical Branch, Galveston, Texas 77550 [81, 151]

L.L. Sackett, Department of Anatomic and Clinical Pathology, St. Paul-Ramsey Medical Center, St. Paul, Minnesota 55101 [165]

Ingrid Sauerbier, Department of Anatomy, Medizinische Hochschule Hannover, Hannover, Federal Republic of Germany [143]

Lawrence E. Scheving, Department of Anatomy, College of Medicine, University of Arkansas for Medical Sciences, Little Rock, Arkansas 72205 [xv, 39]

J. Stutzmann, F.R.A. 15, INSERM, Institut de Physiologie, Faculté de Médecine, 4 rue Kirschleger, 67085 Strasbourg Cedex, France [195]

J. Swoyer, Department of Anatomic and Clinical Pathology, St. Paul-Ramsey Medical Center, St. Paul, Minnesota 55101 [165]

Yasuo Uchiyama, Department of Anatomy, Tohoku University School of Medicine, Sendai 980, Japan [117]

Eleventh International Congress of Anatomy

CONTENTS OF PART A: GLIAL AND NEURONAL CELL BIOLOGY
Sergey Fedoroff, *Editor*

STRUCTURE AND FUNCTION OF ASTROCYTES

Properties of Putative Astrocytes in Colony Cultures of Mouse Neopallium /
   S. Fedoroff, R. White, L. Subrahmanyan, and V.I. Kalnins
Orthogonal Assemblies of Intramembranous Particles — An Attribute of the
   Astrocyte / J.J. Anders and M.W. Brightman
Postnatal Development of Astrocytic Glia in the Cerebellum of *Cyprinus carpio* /
   A. Carrato, A. Toledano, and M.A. Barca
Functional Interactions Between Astrocytes and Neurons / L. Hertz
The Glial Cell as a Major Site of Glycoconjugate Synthesis in the Brain /
   Ch. Pilgrim and I. Reisert
Glial Fibrillary Acidic (GFA) Protein Immunocytochemistry in Development and
   Neuropathology / Lawrence F. Eng and Stephen J. DeArmond

STRUCTURE AND FUNCTION OF OLIGODENDROCYTES

Postnatal Development of Oligodendrocytes / J.A. Paterson
Proliferation of Oligodendroglial Cells in Normal Animals and in a Myelin
   Deficient Mutant-Jimpy / Robert P. Skoff
Remyelination in the CNS / W.F. Blakemore

STRUCTURE AND FUNCTION OF MICROGLIA

Microglia, Monocytes, and Macrophages / Erle K. Adrian, Jr., and
   Robert L. Schelper
Origin of Microglia: Cell Transformation From Blood Monocytes Into
   Macrophagic Ameboid Cells and Microglia / Kikuko Imamoto
Origin, Morphology, and Function of the Microglia / S. Fujita, Y. Tsuchihashi,
   and T. Kitamura

MORPHOLOGICAL ASPECTS OF INTERNEURONAL COMMUNICATION

Quantitation of Developing and Adult Synapses / G. Vrensen and L. Müller
Synaptic Ultrastructure in Unanesthetized and Experimentally Modified Cerebral
   Cortex / D.G. Jones
Plasticity of Synaptic Size With Constancy of Total Synaptic Contact Area on
   Purkinje Cells in the Cerebellum / D.E. Hillman and S. Chen

MORPHOLOGICAL BASIS OF NEUROPHYSIOLOGY OF THE CEREBELLUM

Transmission and Scanning Electron Microscopy and Ultracytochemistry of Vertebrate and Human Cerebellar Cortex / *Orlando J. Castejón and Haydée V. Castejón*
Structure and Fiber Connections of the Cerebellum / *J. Voogd, F. Bigaré, N.M. Gerrits, and E. Marani*
Morphological Correlates of Cerebellar Purkinje Cell Activity / *Jean C. Desclin, F. Colin, and J. Manil*

CONTRIBUTION OF METALLIC IMPREGNATION TO NEUROANATOMY

Neocortical Endeavor: Basic Neuronal Organization in the Cortex of Hedgehog / *F. Valverde and L. López-Mascaraque*
The Golgi-EM Procedure: A Tool to Study Neocortical Interneurons / *A. Fairén, J. DeFelipe, and R. Martinez-Ruiz*
Electron Microscopy of Golgi Impregnated Neurons in Cat Visual Cortex Used to Trace Thalamo-Cortical Projections / *L.J. Garey and J.P. Hornung*
The Golgi Methods and the Abnormal Cerebral Cortex: Epilepsy and Aging / *Arnold B. Scheibel*

DEVELOPMENT OF THE BRAIN

Recent Advances in Child Growth and Development / *Alex F. Roche*

CONTENTS OF PART B: ADVANCES IN THE MORPHOLOGY OF CELLS AND TISSUES
Miguel A. Galina, *Editor*

THREE-DIMENSIONAL MICROANATOMY

Three-Dimensional Microanatomy of Intracellular Structures / *Keiichi Tanaka, Teruyasu Kinose, and Kojiro Atoh*
Morphological Specializations of the Ventricular Surface of the Choroidal Epithelium and Associated Epiplexus Cells / *Delmas J. Allen, G.J. Highison, H. Werneck, and G. Gentry*
Characterization of Free Surface Microprojections on the Kidney Glomerular Epithelium / *Peter M. Andrews*
Surface Morphology of the Papilla and Pelvis in Mammalian Kidneys / *Mario Castellucci*
Optimal Conditions for the Study of Surface Topography of Granulosa Cell Cultures / *Thomas M. Crisp and J. Steven Alexander*
Gross Internal Structure of the Human Kidney / *Gabor Inke*
In Vivo Microscopy of Internal Organs / *Robert S. McCuskey*
The Vascular Anastomoses of the Human Heart / *José António Rebocho Esperança Pina and José Gonçalves Pina*
The Anatomy of Hominization / *Phillip V. Tobias*
Insulo-Acinar Portal System of the Pancreas. A Scanning Electron Microscope Study of Corrosion Casts / *Osamu Ohtani and Tsuneo Fujita*

STEREOLOGY

The Use of Stereology Demonstrated by the Evaluation of Man's Cerebral Cortex / Herbert Haug
Automatic Image Analysers and Their Use in Anatomy / S. Bradbury
Stereological Analysis of the Thyroid Gland by Light Microscopy / Miroslav Kalisnik
Stereology of Colonic Adenomata / Hans Elias, Dallas M. Hyde, Donald S. Mullens, and Fredereck C. Lambert
The Numerical Densities of Alpha and Gamma Motoneurons in Lamina 1X of the Cervical Cord of the Rat: A Method of Determining the Separate Numerical Densities of Two Mixed Populations of Anatomically Similar Cells / Vyvyan Howard, Rhian Lynch, and Laurence Scales
Morphometry of Cortical Dendrites / H.B.M. Uylings, J.G. Parnavelas, and H.L. Walg
The Aging of Cortical Cytoarchitectonics in the Light of Stereological Investigations / H. Haug, G. Knebel, E. Mecke, Ch. Örün, and N.-L. Sass

SOME FUNCTIONAL AND MORPHOLOGICAL ASPECTS OF HAEMATOPOIESIS

Characterization of Hematopoietic Stem Cells. The Fundamental Role of the Transitional Cell Compartment / Joseph M. Yoffey
Transitional and Lymphoid Cells in the Peritoneal Cavity / Joseph M. Yoffey and Pirhiya Yaffe
The Thymus and Haemopoiesis / Marion D. Kendall
Response of Marrow Adipocytes to Hypoxia and Rebound / Bernard G. Slavin, Joseph M. Yoffey, and Pirhiya Yaffe
Stem Cells in the Circulation / Gregor Prindull
Structure and Function of Sinusoidal Endothelium of Bone Marrow / Mehdi Tavassoli
Changes in Bone Marrow Following Sub-Lethal Irradiation / P.F. Harris
The Control of Haematopoietic Stem Cell Proliferation by Humoral Factors / A.C. Riches, M.J. Cork, and D. Brynmor Thomas
The Regulation of the Haematopoietic Stem Cell Compartment in Foetal Liver / D. Brynmor Thomas, M.J. Cork, and A.C. Riches

MORPHOLOGY OF THE RESPIRATORY SYSTEM

Three-Dimensional Electron Microscopy of the Respiratory Tree in the Mammalian Lung / Frank N. Low
Structure of Pulmonary Lymphatics and Lung Clearance / Joseph M. Lauweryns
Role of the Bovine Lymph System in the Defense Mechanism of the Distal Lung / Miguel A. Galina and Robert A. Kainer
Morphophysiologic Bases for the Predisposition of the Bovine Lung to Bronchial Pneumonia / Robert A. Kainer and Donald A. Will

MORPHOLOGY OF THE REPRODUCTIVE SYSTEM

Scanning Electron Microscopy of Normal and Anovulatory Human Ovaries / Sayoko Makabe
3-D Microanatomy of Human Reproductive Organs / E.S.E. Hafez
Gap Junctions in Theca Interna Cells of Developing and Atretic Follicles / G. Familiari, S. Correr, and P.M. Motta
Pineal Control of Reproduction / Russel J. Reiter

Involvement of the Adrenals in Ovulation Induced by Unilateral Ovariectomy in the Rat / *Cl. Aron and J. Roos*
Ultrastructural Differentiation of Leydig Cells in the Testis of 17-Day-Old Chick Embryo and Newly Hatched Chickens / *E. Pedernera, M.C. Aguilar, and M. Romano*
Cellular Interrelationships in the Human Fetal Ovary and Testis / *Bernard Gondos*
Morphogenesis of the Ovary From the Sterile W/W^v Mouse / *H. Merchant-Larios and B. Centeno*

Preface

During the Tenth International Congress of Anatomy held in Tokyo in 1975, it was suggested that a symposium on chronobiology and its relationship to the morphological sciences be organized for the 1980 meeting. In July 1978, Professor Heinz von Mayersbach wrote to Professor Enrique Acosta Vidrio, President of the Eleventh International Congress of Anatomy to be held in Mexico City, proposing that such a symposium be planned. Professor Acosta Vidrio enthusiastically endorsed the idea, and Professor von Mayersbach agreed to be principal organizer.

An outline of the symposium was developed during and immediately after the 1979 meeting of the International Society for Chronobiology in Hannover, Germany. It was decided that the program would deal with chronobiology as it relates to anatomy.

In this volume we present the 13 papers delivered at the symposium. They are supplemented by four others on closely related topics. The first paper, "An Introduction to Chronobiology," is designed to introduce the reader to the field and to stimulate his interest in the articles that follow. These are grouped under five headings: Rhythms in Cells and Tissues, Chronohistochemistry, Rhythms in Susceptibility, Circannual Rhythms, and Control of Rhythms.

Professor von Mayersbach was unable to attend the Congress in Mexico City, and he died before this book was completed. We acknowledge his important role in organizing the symposium and selecting the participants.

Lawrence E. Scheving
John E. Pauly

AN INTRODUCTION TO CHRONOBIOLOGY

John E. Pauly

Department of Anatomy
University of Arkansas for Medical Sciences
Little Rock, Arkansas U.S.A.

The purpose of the paper is to provide a brief introduction to chronobiology and to prepare the reader for some of the more detailed presentations which follow. <u>Chronobiology</u> may be defined as that branch of science which objectively explores and quantifies mechanisms of biological time structure, including the important rhythmic manifestations of life (Halberg et al, 1977). It is a study of biological time structure, that is biological rhythms. Such rhythms are ubiquitous. They are found at all levels of biological organization (Pittendrigh, 1960), from individual cells and their constituents to the behavioral patterns of man.

In addition to introducing the subject, three important questions will be addressed. First, are biological rhythms mere curiosities, important only to chronobiologists, or do they relate to every modern branch of biological science? Second, are the principles of chronobiology simply tools to be used such as a new piece of biomedical instrumentation or a new technique; or are they essential to the design of every experimental protocol, the conduct of the experiment and the interpretation of the results? Finally, are the principles of chronobiology complicated, or are they relatively simple to learn and easy to apply? In short, should all biological science be chronobiology? Should all biological science be related to time? Should all biological scientists be chronobiologists?

BASIC TERMINOLOGY

Like every other branch of science, chronobiology has its own vocabulary. Only those terms and techniques necessary to understand the papers which follow will be discussed here. More complete lists are readily available (Halberg and Katinas, 1973; Halberg et al, 1977).

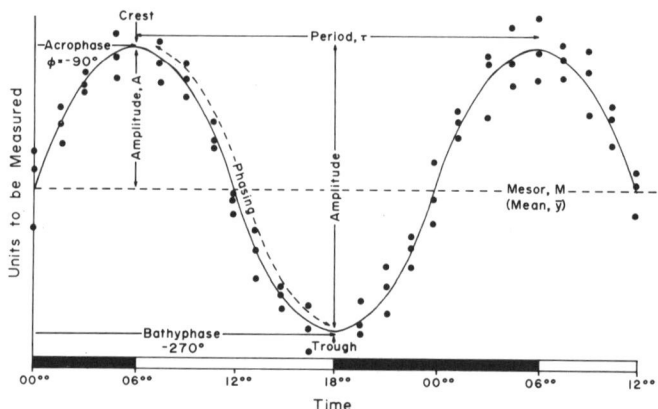

Fig. 1. Chronogram of typical rhythm showing individual data points, phasing or shape, mean, period, two ways of defining its amplitude, two specific phases (peak and trough) and both the acrophase and bathyphase (referenced from local midnight).

Since chronobiology considers biological rhythms, a brief review of the properties of rhythms is appropriate. All rhythms have phasing, a period or frequency, a mean or level, amplitude and phase (Fig. 1). The phasing of a rhythm is its shape. This can be visualized by preparing a graph with the abscissa representing time (milliseconds, minutes, hours, days, months, years, etc.) and the ordinate depicting the units to be measured (milligrams, density, activity, death, etc.), entering the data points, and drawing a line through them. Such a line may be drawn through individual points or through means determined for each time. The period of a rhythm is the time it takes to complete one full cycle. The frequency is simply the reciprocal of period, that is one divided by the period. The mean or level is the average of all the data points along the time scale. The amplitude may be defined two ways. It may be the distance from the crest to the trough or the distance from the mean to the crest. The latter is used by most chronobiologists when employing the cosinor technique to analyze their data (to be described later). The word phase is used to describe a specific point on a rhythm, e.g., the crest or trough.

There are two kinds of time-series sampling, transverse and longitudinal. An example of transverse or cross-sectional sampling would be to measure blood glucose in ten individuals at one time, ten other persons at another time, ten more at still another time, etc. An example of longitudinal sampling would be to measure blood glucose from the same ten individuals at multiple time points along a 24-hour or longer scale. Some biologists employ multiple-time-point sampling, while others persist in using single-time-point sampling. When multiple-time-point sampling is done properly, it may lead to an adequate description of a natural, predictable and regularly-repetitive biological variation. The papers which follow show clearly that single-time-point sampling often leads to erroneous results.

Most of the terms used to this point are common to almost every rhythm, biological or otherwise. There are, however, terms more common to chronobiology than to other disciplines. Many of these relate to periods or the frequency of rhythms (Pauly, 1980). Ultradian rhythms are those with periods shorter than 20 hours. They may vary from the milliseconds of a nerve impulse, to seconds in the respiratory cycle, to the minutes and hours of the human sleep cycle. Circadian rhythms are those with periods ranging from 20-28 hours. Infradian rhythms are those with periods longer than 28 hours. Examples of some of the terms that have been employed to describe these infradian rhythms are circaseptan, circatrigintan and circannual (seasonal). A circaseptan rhythm is one with a period of about one week. Examples of such rhythms may be seen in the excretion patterns of certain urinary steroids (Halberg et al, 1965). A circatrigintan rhythm has a period of about one month. A good example is the human menstrual cycle. A circannual rhythm has a period of about one year, and an example is the reproductive cycle of the deer.

Most of the rhythms that will be described in this symposium are circadian. A <u>circadian rhythm</u> is defined as a regularly repetitive, quantitative physiological change having a period of about 24 hours (circa = about; dies = day; Halberg, 1959). A circadian rhythm is not a diurnal rhythm. The latter term should be reserved to describe a rhythm which occurs during a period of light just as the term nocturnal is used to describe things or events that relate to darkness.

PROPERTIES

Biological rhythms have many properties (Pittendrigh, 1960). To name a few, they are ubiquitous, innate, endogenous, entrainable, self-sustaining, relatively temperature independent, relatively unsusceptible to chemical perturbations and are free-running.

1. <u>Ubiquitous</u>. Rhythms occur at all levels of biological organization, with the possible exception of the prokaryocytes. During this symposium, examples at the cellular level will be provided by Dr. Philippens from Hannover, Germany; Dr. Dalwadi-Bhattacharya from Ahmedabad, India; Dr. Laerum from Bergen, Norway; and Dr. Uchiyama from Sendai, Japan. Dr. Philippens will report on a circadian rhythm in binucleated hepatocytes in rats, and Dr. Uchiyama will discuss circadian variations in lysosomal enzymes in these same cells. At the level of the whole organism, Professor Rietveld from Leiden, The Netherlands, will present evidence that the suprachiasmatic nucleus is responsible for the generation of the circadian rhythm in food intake.

2. <u>Innate</u>. Biological rhythms are not learned; instead they are programmed by the genome (Bünning, 1935). By carefully timing the matings of rats at multiple time points during a 24-hour period and then administering nitrogen mustards at different times of the day and night to dams with pups of identical gestational ages, Dr. Sauerbier from Hannover, Germany, will show that the teratogenic effects are circadian-stage dependent. This important new finding complements the work of others who have demonstrated the hours of changing resistance and susceptibility to many drugs (Carlsson and Serin, 1950a, 1950b; Haus and Halberg, 1959; Pauly and Scheving, 1964; von Mayersbach, 1974; Scheving et al, 1974; Scheving et al, 1980); however she is

the first person to prove that even embryos of identical age are more susceptible to an insult at one circadian stage than at another.

3. <u>Endogenous</u>. Rhythms are controlled within the cell or organism. Although Professor Brown at Northwestern University in Evanston, Illinois, U.S.A., and his colleagues persist in the view that rhythms are controlled by exogenous geophysical forces such as magnetism or cosmic radiation (Brown, 1980), the majority of chronobiologists hold to the view that the natural periods of rhythms will continue in the absence of any environmental forces or time cues. Even though rhythms are innate, certain exogenous or endogenous factors can influence them to a considerable degree. Professor Scheving from Little Rock, Arkansas, will demonstrate such an influence by describing mitotic rhythms in gut epithelium and the effect of epidermal growth factor on them.

4. <u>Entrainable</u>. Environmental factors called <u>synchronizers</u>, Zeitgebers, entraining agents or time cues synchronize circadian rhythms to exact 24-hour periods (Halberg, 1969). Light is the dominant synchronizer for most plants and animals, and the alternating periods of light and darkness resulting from the rotation of the earth are responsible for the maintenance of the precise 24-hour periods that characterize so many of the rhythms in nature. Whereas the dominant synchronizer for the rodent is light, in man it is the societal regimen (Scheving and Pauly, 1977). It will be demonstrated later in this paper, that totally blind people maintain precise 24-hour circadian rhythms as long as they continue to interact with other human beings who follow regular routines (Pauly et al, 1975). As long as the synchronizer is applied once every 24 hours, the organism will be entrained to the exact period of the solar day. However, in the case of many animals, the periods of light also will influence the rhythm. Dr. Oudet from Strasbourg, France, will first demonstrate both the circadian and circannual rhythms in mitosis in the epiphyseal cartilage of rats and then show that growth rate can be increased by lengthening the periods of light to which they are subjected. It takes rhythms varying lengths of time to synchronize to a new lighting regime. Dr. Dalwadi-Bhattacharya will report that eight days after reversing the lighting from LD to DL on a colony of rats, the number of lysosomes and the intensity of staining for acid phosphatase, β glucuronidase and arylesterase was altered; however the

three enzymes differed in their adaptive capacities as evidenced by differences in the intensities of their reactions.

5. <u>Self-Sustaining</u>. Rhythms continue in the absence of a synchronizer. They do not dampen when the organism is removed from time cues and placed in constant light (LL) or constant darkness (DD).

6. <u>Relatively Temperature Independent</u>. Although Dr. Oudet will show that growth rate in epiphyseal cartilage also can be increased by maintaining experimental rats for long periods in a low-temperature environment, the periods of rhythms usually are unaffected by temperatures within the range the organism can tolerate. Animals also are relatively unsusceptible to chemical perturbations; however Miss Sackett from Minneapolis, Minnesota, U.S.A., will show that a mitotic rhythm in a murine melanoma can be synchronized with hydroxyurea at one circadian stage but not at another.

7. <u>Free-Run</u>. In the absence of any synchronizer, that is in constant conditions, a rhythm that has been operating on an exact 24-hour basis will revert to its natural period. If such a period happened to be 25 hours in length, then a week after the constant conditions were imposed, an animal would be 7 hours out of phase with others that had been maintained under the original conditions. If the crest of any particular rhythm operating within the organism had originally occurred at 12 midnight local time, after one week in the constant conditions, it would be expected to occur at 7:00 in the morning.

The ability to free-run is very important to plants and animals in their natural environment. As periods of light and darkness change with the seasons, the organisms are able to adjust themselves accordingly by free-running. The changing photoperiods are responsible for the induction of hibernation, migratory behavior, annual reproductive cycles, etc. It has been suggested that the <u>purpose of biological rhythmicity</u> is so that organisms can adapt themselves in advance of changes that will occur in the environment (Halberg, 1969).

EXAMPLES OF BIOLOGICAL RHYTHMS

The human temperature rhythm has been known since 1873; however there are some who believe that the rise in temperature which occurs in the morning and persists throughout the day is attributable only to an increase in bodily activity or the consumption of food. Figure 2 illustrates the rectal temperature of a subject over two 24-hour periods while he remained continuously resting in bed. A very regular rhythm is evident with a crest in the late afternoon and a trough in the early morning hours. The subject fasted during the second 24-hour period, and it is apparent that the lack of food had no effect whatsoever on the rhythm in body temperature. This is a perfect example of a circadian rhythm, that is, a regularly repetitive, quantitative, physiological change having a period of about 24 hours. In the case of man, this rhythm is synchronized to an exact 24-hour period by the societal regimen.

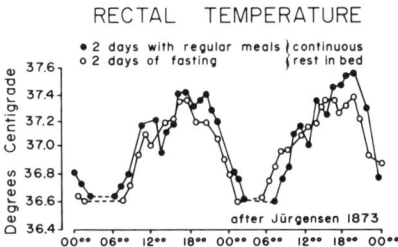

Fig. 2. Rectal temperature of man continuously in bed for 48 hrs. Regular diet first 24 hrs.; fasting second 24 hrs.

Since rhythms are regularly repetitive and can be synchronized by factors in the environment, it follows that organisms maintained under similar conditions and appropriately synchronized should have similar phasing in their rhythms. A demonstration of this supposition is apparent in Figure 3 which depicts two studies on the mitotic rhythm in human epidermis. The first was done in 1959 and the second in 1968. The phasing of the two rhythms is almost identical, and the differences in amplitude possibly might be explained on the basis of different populations of subjects, different investigators doing the mitotic counts or even the overriding influence of a circannual rhythm. The point to be made, however, is the remarkable similarity in the results of the two studies done nine years apart by different investigators, in different locations on separate populations.

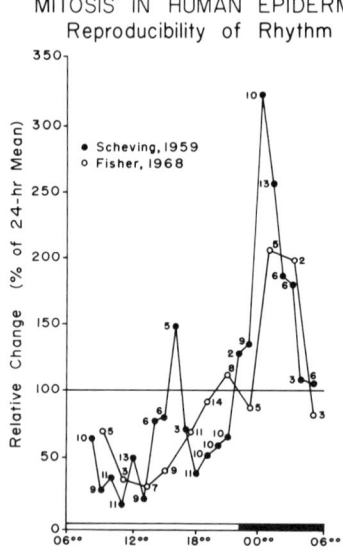

Fig. 3. Rhythm of mitosis in human skin. Results of two studies, one by Scheving in 1959 and the other by Fisher in 1968.

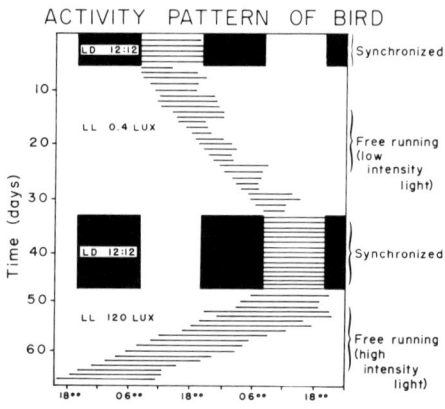

Fig. 4. Activity pattern of single bird maintained first in alternating periods of light and darkness, second in constant dim light, third in alternating periods of light and darkness and fourth in constant bright light. (Modified after Aschoff.)

By definition a true circadian rhythm has a period of <u>about</u> 24 hours. In constant conditions, without time cues, its period will deviate from the exact 24-hour day by a fixed amount each day, and the rhythm is said to free-run. An example of free-running may be seen in Figure 4 which depicts the activity pattern of a bird first maintained in alternating 12-hour periods of light and darkness; second, in constant dim light; third, again in alternating periods of light and darkness; and finally in constant bright light. While subjected to alternating 12-hour periods of light and darkness, the bird was active during the light phase and quiet in the dark. In constant conditions, without time cues from the environment, its activity pattern free-ran. In conditions of constant dim light, its activity began a little later each day, which means the period was slightly longer than 24 hours. After about 20 days it can be seen that the bird started its activity at about the same clock time that it had started its rest period at the beginning of the experiment. In the third phase of the study the bird once again was subjected to alternating periods of light and darkness and immediately synchronized its activity to the periods of light. In the final phase, the bird was subjected to constant bright light. Once again its rhythm deviated from the 24-hour day by a fixed amount and began to free-run. This time, however, under the conditions of bright light, the period of the rhythm was shorter than 24 hours, that is, the bird began its activity a little earlier each day. Thus we see that not only is light a dominant synchronizer of most plants and animals, but its intensity can influence the period of the rhythm under certain circumstances. These facts are summarized in what is popularly known as the Circadian Rule which states that with increasing intensities of illumination, the circadian period is shortened in diurnal (light-active) animals and lengthened in nocturnal (dark-active) animals (Aschoff, 1952). Thus in a bird, the period in constant light is less than it is in constant darkness. Figure 4 illustrates this point by showing that the period of the bird's activity-rest rhythm was more than 24 hours in the very dim light but less than 24 hours in bright light. Figure 4 illustrates at least three properties of rhythms: they are endogenous, entrainable by forces in the environment, and the intensity of light may influence their periods.

The rhythms of man will free-run in a manner similar to those in experimental animals. When a person flies in a jet plane from Chicago to London non-stop, six time zones are crossed within a relatively short period of time. Upon arrival in London, the individual is immediately subjected to a new societal routine, that is, a new set of environmental synchronizers. For varying periods of time his biological rhythms are said to be out of phase with their synchronizers, and the man usually suffers from a condition popularly known as "jet lag". This condition or disease is slowly corrected during the next seven or eight days as the circadian rhythms free-run until they entrain to the new synchronizer (societal routine). Many biological rhythms operate within a single organism. Although some of these rhythms have precise phase relationships with each other and with the environmental synchronizers, they may have different natural periods. Therefore when they free-run after a rapid transmeridianal displacement, some rhythms re-establish a correct phase relationship to their synchronizer faster than others. It may take only two or three days for the activity-rest rhythm to adjust but two or three weeks for the temperature rhythm. As a consequence, an individual may not feel completely well until all of his or her body rhythms have been resynchronized to the new environment. In a study of pilots flying commercial jet aircraft between Germany and the United States, Klein et al (1970) found that the rate of phase adjustment was approximately 1.5 (1-2) hours each day. It took the pilots about five days to resynchronize to their new environments after an eight-hour time shift.

Not all animals respond the same way to environmental influences. For example, the free-running activity period of the nocturnal wild rat will increase in constant bright light, whereas that of the diurnal bird will decrease under these same circumstances. On the other hand, a change in light intensity has relatively little effect on diurnal man.

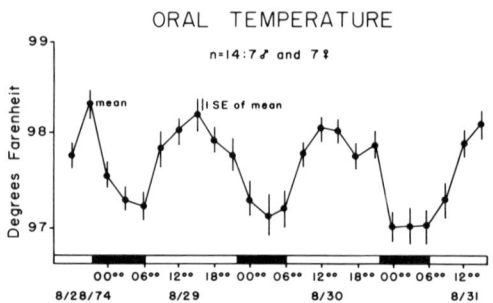

Fig. 5. Oral temperature in 14 blind humans.

It has been suggested that the rhythms of blind people are different from those of sighted individuals. Figure 5 is a chronogram depicting the rhythm of oral temperature in

14 blind subjects. The phasing, mean, period and amplitude are identical to those seen in sighted people. However when one compares the rhythm in urinary epinephrine of blind people with that for people with normal sight, some striking differences are seen (Fig. 6). The mean and amplitude are much higher in the blind people, but the period and phasing are similar in the two populations.

For many years it has been known that cortisol levels in humans are very high in the early morning hours but decline throughout the day and early evening only to rise again before we awaken the next day. When the rhythm of human cortisol is drawn on the same graph with the rhythm of corticosterone of the rat, it is seen that cortisol peaks at the same time corticosterone is in a trough phase and vice versa (Fig. 7). This might be expected when one considers that man is a diurnal animal and rats are nocturnal. Based on this single finding, one might speculate that all the rhythms of man and rodents are 180° out of phase, but this simply is not true. Many rhythms in both species have been studied with great care; some are very similar, and others like those of cortisol and corticosterone are as much as 180° out-of-phase.

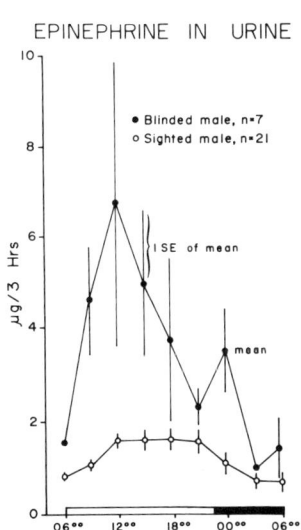

Fig. 6. Comparison of epinephrine in urine of blind and sighted men.

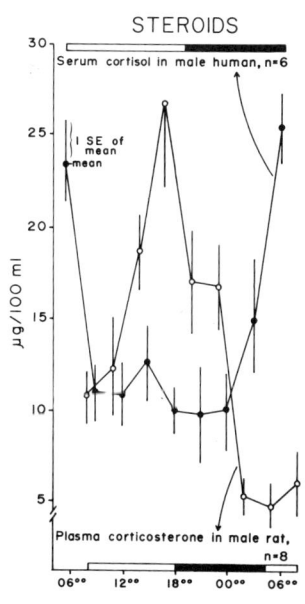

Fig. 7. Comparison of serum cortisol in diurnal man with plasma corticosterone in nocturnal rats.

Among the many rhythms that have been characterized in the human is one for the enzyme amylase in saliva. The question might be asked whether such a rhythm is dependent upon the diet. Does the time when meals are eaten influence the rhythm? Figure 8 compares the rhythms in amylase in persons on a controlled diet (fixed amounts of corned beef hash and distilled water every three hours throughout the 24-hour span) with that in individuals who are fasting. No differences in the rhythms of the two populations are seen.

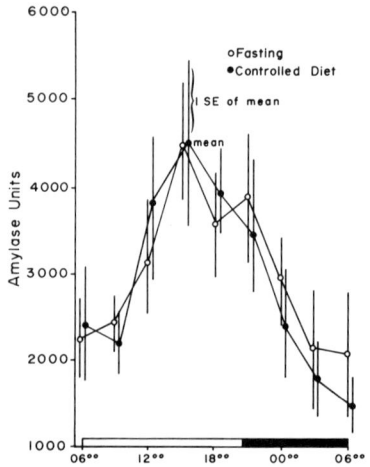
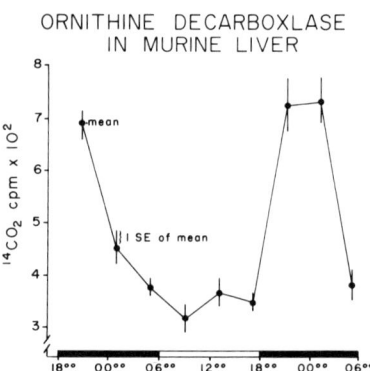

Fig. 8. Comparison of rhythm of amylase in saliva in subjects on controlled diet and during fasting.

Fig. 9. Circadian variation of ornithine decarboxylase activity in murine liver.

Ornithine decarboxylase is an enzyme that has been closely related to cell replication. In the liver of the mouse, a peak in ornithine decarboxylase activity occurs at night and a trough during the daylight hours (Fig. 9). The peak corresponds to the time when there is a great deal of DNA synthesis. A peak in mitosis occurs at the same time the level of ornithine decarboxylase is low. Later during the symposium Dr. Philippens will report on a rhythm in hepatic glycogen for Professor von Mayersbach of the Medizinische Hochschule in Hannover, Germany.

There is a well-established rhythm in the incorporation of tritiated thymidine into DNA in the esophagus of the mouse (Fig. 10). The highest counts coincide with the peak of the synthesis phase of the cell cycle. In some tissues there is a high level of DNA synthèsis at all times; in others very little occurs at one time along the 24-hour scale, whereas at another there is a great deal of synthesis in preparation for the next wave of mitotic activity. A peak in DNA synthesis does not necessarily occur at the same time in different organs. For example, in the esophagus the peak occurs late in the dark phase and the trough at the end of the light phase. Using the new flow cytometric method, Dr. Laerum from Bergen, Norway, will show the relative distribution of DNA in bone marrow and peripheral blood and prove that cell renewal in bone marrow is circadian-stage dependent.

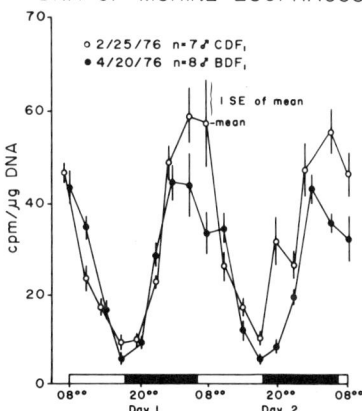

Fig. 10. Circadian rhythm in the incorporation of tritiated thymidine into DNA of mouse esophagus. Results of two studies.

A question that often is asked is what is the purpose of the circadian system? Halberg (1969) has suggested that it provides a base or mechanism for the adjustment of the organism to the periodic changes in the environment. It adjusts the animal continuously and in advance to circumstances that are to be expected or experienced in the near future. For example the level of cortisol rises before one gets up in the morning, perhaps in anticipation of the activity expected during the day (Fig. 7).

ANALYSIS OF THE DATA

Chronobiologists use many techniques to analyze their data. Time-series data often are difficult to accumulate at equidistant points; this is particularly true when collecting specimens from humans under clinical conditions. A very powerful inferential statistical technique that lends itself nicely to the analysis of such data is commonly called the "cosinor" method (Halberg et al, 1967, 1972; Nelson et al, 1979). This analysis (performed readily by a computer) will objectively determine a probability or P value, which will indicate the significance of a fit of the data to a cosine curve by least squares (if the P value is 0.05 or less, the function of the variables studied is presumed to be cyclic and not random), and three rhythmic parameters and their dispersions: the acrophase (ϕ), mesor (M) and amplitude (A). The acrophase represents the crest of the fitted cosine curve in relation to some arbitrarily selected reference point along the 24-hour time scale. Usually the acrophase corresponds to the time when the data values are, on the average, highest; however, it should be noted that the acrophase is not necessarily the time when the peak value was recorded. The reference point chosen for many studies is mid-sleep; but it may be midnight, the time of peak temperature or any other point. Frequently the acrophase is expressed in degrees ($^\circ$) rather than hours. If 360° equal 24 hours, then 15° equal one hour. Thus if the reference point were local midnight, -15° would represent 0100, or 0116 would be -19°. The minus sign preceding the degrees is in keeping with mathematical convention.

The mesor (M) is the cosinor-determined, over-all 24-hour mean; this is equivalent to the 24-hour arithmetical mean only if the data points are equidistant. The amplitude (A) is defined as one-half the total cosine excursion best approximating the rhythm; it represents the distance between the mesor and the crest of the cosine function. (The reader is reminded that the amplitude of the cosine function is not the same as the amplitude of a rhythm that has been plotted directly from the data, in which case the amplitude frequently is expressed as the distance between the trough and the crest; Fig. 1.) In a cosine analysis, both the amplitude and mesor are expressed in the original units of the variables analyzed, i.e., degrees of temperature, micrograms/three hours, etc. The computer can be connected to a plotter and programmed to display the cosinor as a polar

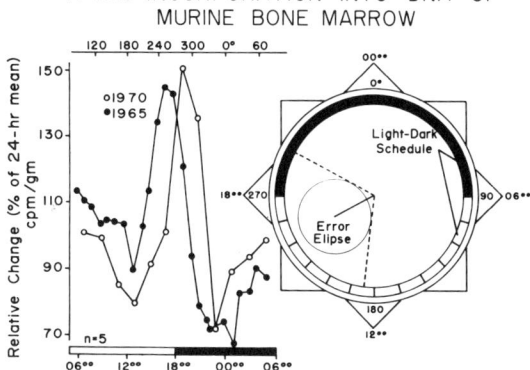

Fig. 11. Circadian rhythm in the incorporation of tritiated thymidine into DNA of rat bone marrow. Comparison of chronogram of two studies and single cosinor analysis.

plot (Fig. 11). The length of the line drawn from the pole towards the periphery represents the amplitude of the rhythm. The line itself points to the acrophase, that is the lag time from the selected reference point to the crest of the computer-derived cosine function. The statistical error ellipse depicts the region of 95% confidence, and lines drawn from the pole through the edges of the error ellipse to the periphery provide the 95% confidence arc. Such a display is particularly useful when comparing two or more rhythmic functions on the same graph. As long as the error ellipses do not overlap the pole or each other, the acrophases may be assumed to be statistically different. A comparison of a chronogram and a cosinor analysis depicted as a polar plot can be made in Figure 11. The chronogram on the left shows the results of two experiments done to measure the uptake of tritiated thymidine into bone marrow at multiple times points along a 24-hour scale. The first experiment was performed in 1965 and the second in 1970. There is a remarkable similarity in their means, amplitude and phasing. On the right the data are displayed as a polar plot of a single cosinor analysis. The size of the error ellipse, its proximity to the pole and the wide confidence arc correspond to a P value of .021.

The cosinor technique can be very useful when comparing two or more functions simultaneously in an organism living under various experimental conditions. For a period of more than four months, a woman monitored her rectal temperature and collected her urine at precise periodic intervals for analyses of 17-hydroxycorticosteroids. She started accumulating data while she followed her usual 24-hour societal routine, continued while she lived for almost three months in a cave free from all time cues, and then extended her collections for another month, after she emerged from isolation and once again pursued a normal routine. All her data were analysed by the cosinor technique, and the acrophases of both physiological functions were plotted along a time scale covering more than four months (Fig. 12). The rhythms of both temperature and steroid excretion changed their periods during isolation and free-ran, but they continued to maintain their phase relationship to each other. The acrophase of the temperature rhythm lagged behind that of steroid excretion before and during isolation as well as during the period of resynchronization with the societal routine.

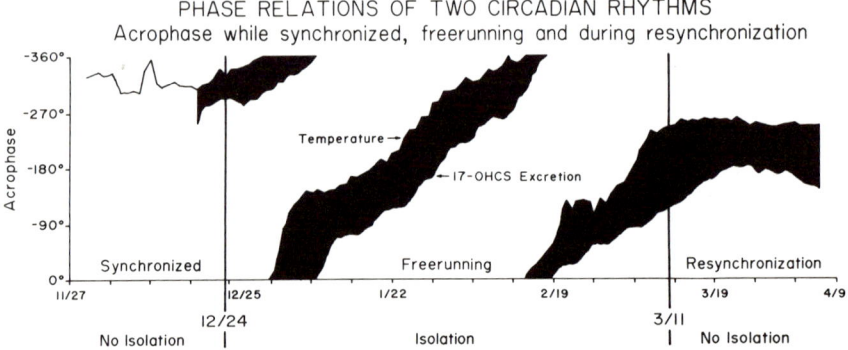

Fig. 12. Acrophases of circadian rhythms of rectal temperature and urinary excretion of 17-hydroxycorticosteroids in a woman before, during and after a three month period of isolation in a cave. Both functions free-run but maintain proper phase relations throughout study. (Modified after Halberg)

Professor Scheving will show how an acrophase map was constructed following cosinor analyses of the data from a study which involved multiple variables. On one chart will be depicted the acrophases and confidence limits of a host of physiological variables in human vital signs and the constituents of serum and urine in a group of 13 young soldiers. Each of the variables measured has its own particular rhythm. Although some of these rhythms have specific phase relationships with each other, their acrophases occur at all different times of the day and night. By studying such an illustration, it becomes very obvious that the human organism is never in a homeostatic or constant state; rather life should be redefined in terms of multiple, independent, phase-related and integrated physiological, rhythmic functions that make us different anatomically, biochemically and physiologically at various times of the day, the month and the year.

PRACTICAL APPLICATIONS

The fact that we are different biological entities at different times has enormous practical significance. For example, if a fixed dose of a drug based on body weight is given to subgroups of animals at multiple time points along the 24-hour scale, one may expect different responses at different times. A dose of pentobarbital sodium that will cause a rat to sleep less than an hour at one time of the day will produce sleep for about an hour and a half at another time (Fig. 13). When the dose of the drug is raised to the so-called LD:50 level, much more dramatic results may be expected. Relatively few animals will succumb to the toxicity at one time of the day, but as many as 90% of them will die at another time (Pauly and Scheving, 1964). Thus it becomes apparent that the time of administration may be a matter of life or death depending upon the circadian phase of the organism. Professor Scheving will elaborate on this point in his paper. Dr. Rubin from Galveston, Texas, U.S.A., will first show a dose response to ionizing radiation on murine corneal epithelium *in vivo* and then prove that the response to a fixed dose of radiation is strongly circadian-phase dependent.

Consideration of biological rhythmicity always is important in the design and execution of experimental procedures and the interpretation of their results. In fact,

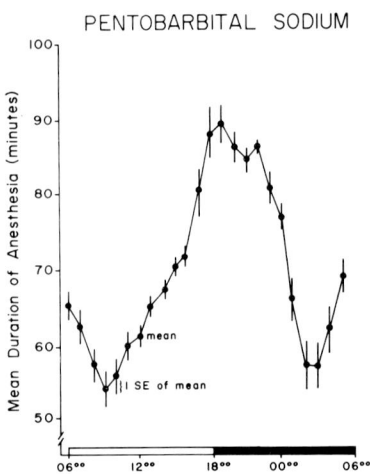
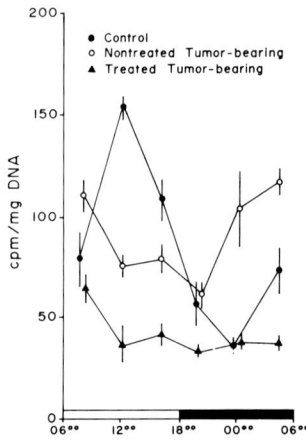

Fig. 13. Duration of anesthesia from fixed dose of pentobarbital sodium at different circadian stages.

Fig. 14. DNA synthesis in rectum in control, non-treated tumor bearing and tumor bearing mice treated with cytosine arabinoside.

in many instances, failure to consider chronobiological implications will result in erroneous conclusions. To illustrate the point, let us assume that we are interested in the effect a chemotherapeutic agent might have on the epithelium of the gut in three populations of animals: untreated control animals without the tumor, untreated animals with a tumor and tumor-bearing animals treated with a course of cytosine arabinoside (Fig. 14). If we decided to sample the three populations at 7:00 or 8:00 in the morning, we probably would conclude that there was little or no difference in DNA synthesis between the control and non-treated tumor bearing populations, but DNA synthesis was elevated in the group that received chemotherapy. On the other hand, if we decided to sample at noon or in the early afternoon, we probably would conclude that DNA synthesis was greatly reduced in tumor-bearing animals, particularly if they were receiving cytosine arabinoside. Finally, if we decided to sample at 7:00 or 8:00 in the evening, we would conclude that there was no difference between the controls and the non-treated tumor-bearing animals, but DNA synthesis was reduced in the tumor-bearing animals receiving the drug. In short, the results of the experiment would depend upon the single time point chosen to do the sampling. Almost certainly we would have come to erroneous conclusions unless we had designed and conducted the experiment so that samples were taken at multiple time points along the 24-hour scale. Thus it is not sufficient to say that circadian variability is not important as long as you sample the controls and experimental animals at the same time of the day. Although such an attitude seems to be prevalent among persons with limited knowledge of chronobiology who do not wish to be "bothered" with the problem of multiple time-point sampling, it often leads to pitfalls, confusion and outright errors.

CONCLUSION

In this brief introduction to chronobiology, it has been demonstrated that biological rhythms are much more than mere curiosities. They relate to all branches of modern biological science. Chronobiology is more than a tool or technique. Its principles are essential to the design, conduct and interpretation of biological experiments.

The basic principles of chronobiology are not particularly complicated. They are relatively easy to apply and insure an extra measure of validity to the results and conclusions of experiments.

It is apparent that all biological science should be related to time, that all biological science really is chronobiology and that we all should be chronobiologists. The remaining papers presented in this symposium will substantiate these claims.

REFERENCES

Aschoff J (1952). Frequenzänderung der Aktivitätsperiodik bei Mäusen im Dauerdunkel und Dauerlicht. Pfluger's Arch 255:197.
Brown FA Jr (1980). The exogenous nature of rhythms. In Scheving LE, Halberg F (eds): "Chronobiology: Principles and Applications to Shifts in Schedules", Alphen aan den Rijn, The Netherlands: Sijthoff & Noordhoff, p 127.
Bünning E (1935). Zur Kenntnis der erblichen Tagesperiodizität bei den Primärblättern von Phaseolus multiflorus. Jb wiss Bot 81:411.
Carlsson A, Serin F (1950a). Time of day as a factor influencing the toxicity of nikethamide. Acta Pharmacol 6: 181.
Carlsson A, Serin F (1959b). The toxicity of nikethamide at different times of the day. Acta Pharmacol 6:187.
Halberg F (1959). Physiologic 24-hour periodicity; general and procedural consideration with reference to the adrenal cycle. Z Vitamin-Hormon-u Fermentforsch. 10:225.
Halberg F, Engeli M, Hamburger C, Hillman D (1965). Spectral resolution in low-frequency, small amplitude rhythms in excreted 17-ketosteroids; probable androgen-induced ciraseptan desynchronization. Acta Endocrinology: suppl. 103, Periodicza Copenhagen, p 5.
Halberg F, Tong YL, Johnson EA (1967). Circadian system phase – an aspect of temporal morphology procedures and illustrative examples. In Mayersbach Hv (ed): "The Cellular Aspects of Biorhythms", New York: Springer-Verlag, p 20.
Halberg F (1969). Chronobiology. Ann Rev Physiol 31:675.
Halberg F, Johnson EA, Nelson W, Runge W, Sothern R (1972). Autorhythmometry procedures for physiologic self-measurements and their analysis. Physiology Teacher 1:1.
Halberg F, Katinas GS (1973). Chronobiologic glossary of the International Society for the Study of Biologic Rhythms. Int J Chronobiology 1:31.
Halberg F, Carandente F, Cornelissen G, Katinas GS (1977). Glossary of chronobiology. Chronobiologia, Vol IV, suppl p 50.
Haus E, Halberg F (1959). 24-hour rhythm in susceptibility of C mice to a toxic dose of ethanol. J Appl Physiol 14:878.

Klein KE, Brüner H, Holtmann H, Rehme H, Stolze J, Steinhoff WD, Wegmann (1970). Circadian rhythm of pilot's efficiency and effects of multiple time zone travel. Aerospace Med 41:125.

Mayersbach Hv (1974). Circadian liver detoxification and acetylcholinesterase rhythmicity: two limiting factors in circadian E600 toxicity. In Scheving LE, Halberg F, Pauly JE (eds): "Chronobiology," Tokyo, Igaku Shoin Ltd, p 191.

Nelson W, Tong YL, Lee JK, Halberg F (1979). Methods for cosinor-rhythmometry. Chronobiologia 6:305.

Pauly JE, Scheving LE (1964). Temporal variations in the susceptibility of white rats to pentobarbital sodium and tremorine. Internat J Neuropharmacol 3:651.

Pauly JE, Scheving LE, Burns ER, Landon J, Stone JE (1975). Studies of the circadian system in blind human beings. In "Proceedings of the XII International Conference", International Society for Chronobiology, Milano, The Publishing House "Il Ponte", p 19.

Pauly JE (1980). The spectrum of rhythms. In Scheving LE, Halberg F (eds): "Chronobiology: Principles and Applications to Shifts in Schedules", Alphen aan den Rijn, The Netherlands, Sijthoff & Noordhoff, p 33.

Pittendrigh CS (1960). Circadian rhythms and the circadian organization of living systems. In: "Biological Clocks", Cold Spring Harbor Symposia on Quantitative Biology, Vol XXV, New York, Cold Spring Harbor, L. I., p 159.

Scheving LE, Mayersbach Hv, Pauly JE (1974). An overview of chronopharmacology. J Europeen Toxicologie 7:203.

Scheving LE, Pauly JE (1977). Several problems associated with the conduct of chronobiological research. In Scharf J-H, Mayersbach Hv (eds): "Die Zeit und das Leben - Chronobiologie", Halle, Nova Acta Leopoldina, p 237.

Scheving LE, Burns ER, Pauly JE, Halberg F (1980). Circadian bioperiodic response of mice bearing advanced L1210 leukemia to combination therapy with adriamycin and cyclophosphamide. Cancer Res 40:1511.

RHYTHMS IN CELLS AND TISSUES

CHRONOMORPHOLOGY OF MAMMALIAN TISSUES

H. v. Mayersbach, K. M. H. Philippens

Department of Anatomy, Med. Hochschule,
Hannover, W. Germany

INTRODUCTION

Since the chronobiochemical and histochemical studies of Forsgren (7) and Holmgren (11), an overwhelming volume of data has demonstrated the essentially rhythmic nature of cell activity in the living organism.
Up to the present, however, in most chronobiological studies reported in the literature, circadian (ca. 24-hour) rhythmic variations have been traced mainly at the level of physiological and biochemical activities; correspondingly the number of chronomorphological investigations is rather few. The main reasons for this are: 1, techniques used in morphology are relatively laborious and time-consuming; this is especially of great disadvantage when investigating multiple samples in time series studies; 2, many investigators did and sometimes still do believe that functional rhythms in cells and tissues are just neglectible minor fluctuations which are not parallelled by visible changes at the structural level; 3, to a certain degree the latter misconception often gained support from technical and statistical problems to quantify and evaluate complex structural configurations as well as differences in intensity of histochemical stainings.
In this outline a selection of well-documented data will be presented in order to demonstrate that: 1, rhythms in cell work can be traced by histochemical techniques visualizing the spatial distribution patterns of substances and enzyme activities; 2, many rhythmic changes at the functional level are ultimately related to and based on structural rearrangements in the cell; 3, consideration of chronobiological concepts and facts is an indispensable precondition for the in-

terpretation of environmental and experimentally induced influences on the animal system.

EXPERIMENTAL DATA AND DISCUSSION

Rhythms Detected by Histochemistry

Prompted by our own experience with the rhythmic fluctuation of glycogen in the liver (16) we carried out systematic chronohistochemical investigations on a number of liver enzymes in normal ad-lib. fed rats and mice. As a result of this we found that the circadian varying content and intralobular distribution pattern of glycogen is parallelled by corresponding variations of glycogen-metabolizing enzyme activities (fig. 1). Furthermore, the activity of several other enzymes related to different metabolic pathways and/or subcellular compartments also revealed similar changes in histo- respectively cytotopochemical distribution and/or reaction intensity; for example, succinic dehydrogenase, glucose-6-phosphatase, 5'-nucleotidase, ß-glucuronidase, indoxylacetat esterase (8,12, 17,19,23). The quantitative variations of some of these activities could be established by parallel biochemical analysis of the same livers.

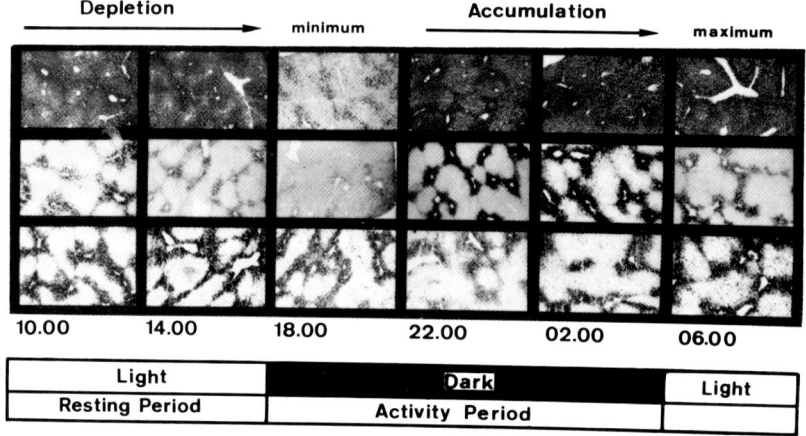

Fig. 1: Circadian rhythmic variations of glycogen (upper row), glycogen synthetase (E.C. 2.4.1.11) (middle) and glycogen phosphorylase (E.C. 2.4.1.1) (lower row) of male NMRI-mouse liver; ad lib. feeding (from 20).

During the last two decades the number of chronohistochemical reports, not only dealing with the liver, but also with highly specialized systems in other organs, e. g., the stomach, intestine and pancreas, has remarkably increased (13,29,33,41). This certainly signalizes the growing awareness of the value of histochemistry as an approach to chronobiology.
On the other hand, the use of daytime-qualified protocols may prove to be a powerful tool to answer basic problems encountered in histochemistry. Thus, histochemical evidence could be given that in rat hepatocytes at certain times of the day 5'Br,4'Cl-indoxylacetate is hydrolyzed only by lysosomal A-esterase, whereas at other timepoints its enzymatic hydrolysis is demonstrable also outside the lysosomes ('cytoplasmic background staining') (14,17). According to current concepts based on time-indifferent studies, A-esterase activity is confined to the lysosomes only, and the dual localization is caused, most probably, by the co-reaction of cytoplasmic B-esterase. However, even after inhibition of E600-sensitive B-esterase, the dual histochemical reaction appears in a daytime-dependent fashion. This phenomenon strongly favours the real periodic occurrence of A-esterase activity in the endoplasmic reticulum which is discussed as one of the sites of lysosome cytogenesis (5,10). At the same time, these chronohistochemical variations may help to elucidate certain problems concerning the use of so-called lysosome marker enzymes, and indirectly, also those problems related to the heterogeneity of liver cell lysosomes.

Rhythmic Variations at the Structural Level.

In rat liver paraffin sections a systematic rhythmic variation of the number of <u>hepatocyte nuclei per microscopical field</u> has been reported by several investigators (6,27,40). Basically, this phenomenon reflects the alternating increase and decrease of liver cell volume during stages of nocturnal substrate storage and of diurnal depletion, respectively, in close correlation to the rat's sleep-wake and feeding cycle. The changes in number of nuclei and in cell volume may strongly influence the result and interpretation of quantitative biochemical determinations of cytoplasmic and, especially, of nucleic substances when performed per unit weight of wet liver tissue. Neglect of this may lead to over- or underestimation, e. g., of changes in DNA concentration (µg/100 mg wet wt) observed in non-rhythmic or in rhythmic studies (27,31).
In direct relation and parallel to the rhythmic changes in he-

patocyte volume, the total wet weight of the liver of rats and
mice varies statistically significant along the 24-hour time
scale (6,27,40). Due to the rhythmic variability of liver protein and DNA, problems may arise as to the usefulness of these
substances as reference values in quantitative liver biochemistry. For this reason, the concentrations or activities of
some liver constituents should preferably be calculated per
total-liver fresh weight or per gram body weight, which is an
appopriate unit in terms of so-called 'bio-availability' (15).
Of course, what is said about the liver basically holds true
also for some other organs, for example the salivary glands
which reveal prominent circadian variations in the size of
their cells and in total mass.

The occurrence of a high-amplitude circadian rhythm in the
relative number of binucleate hepatocytes in rats will be discussed by Philippens elsewhere in this symposium. It throws a
new light on the high responsiveness of the liver-cell nucleus
to physiologic changes of metabolic load during the 24-hour
day. It should be mentioned that the short- or long-term

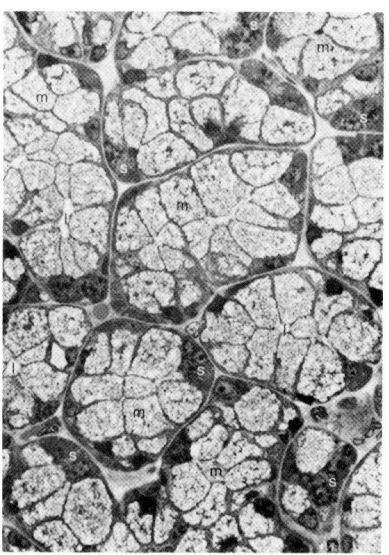 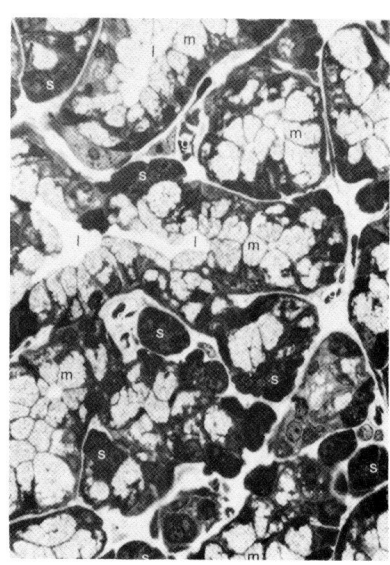

Fig. 2: Sublingual gland in the normal ad lib.-fed male rat
(Semi-thin, toluidin blue-stained sections, magn. 700 x) 1h before the onset at the dark span (left) respectively the light
span (right). m = mucous secretory cell, s = serous secretory
cell, l = lumen of secretory unit, g = blood vessel (from 1).

changes in binuclearity reported to be induced by experimental influences such as drugs, hunger, protein deprivation etc. (3, 21) may be within the range of the spontaneous fluctuation found in rhythmic studies. This observation clearly stresses the necessity of considering chronobiological facts in morphological research.

The 24-hour periodicity of cell division is one of the most extensively investigated rhythmic functions, which can be easily recorded in routine histological sections of several different organs and tissues. In normally growing and regenerating tissues of rodents and other organisms, the mitotic rate systematically fluctuates with maximal values occurring at certain times of the day (e. g. during the animals resting phase) and minimal values about 12 hours later (34,36). Based on the knowledge of this, chronobiological concepts have been successfully applied in combined morphological and nonmorphological studies on the time relationship between different rhythmic variables. For example, in the mouse such relationships have been described between endocrine (adrenocortical pituitary) functions on one hand and biochemical (e. g. energy

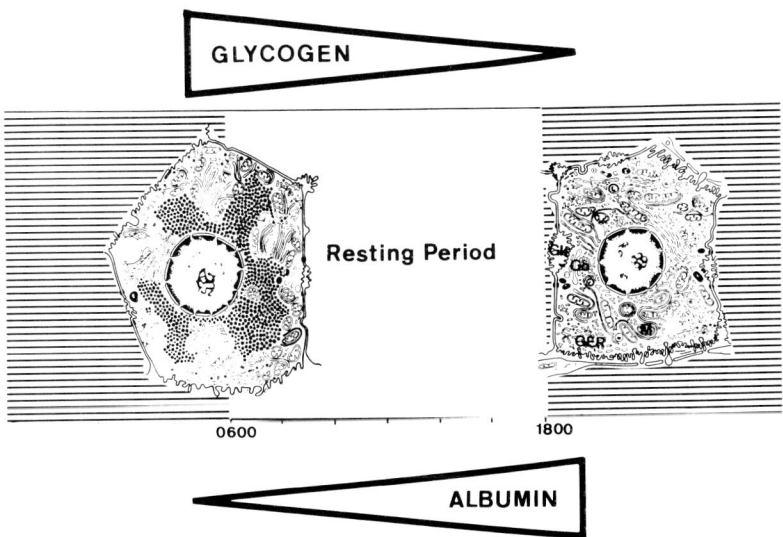

Fig. 3: Rhythmic variations in the ultrastructural organization of normal rat liver parenchymal cells; ad lib.-feeding. Schematic presentation of electronmicroscopic observations by Müller (19).

storage and release) as well as cytokinetic processes in the liver on the other hand (2). The potential meaning of estab-

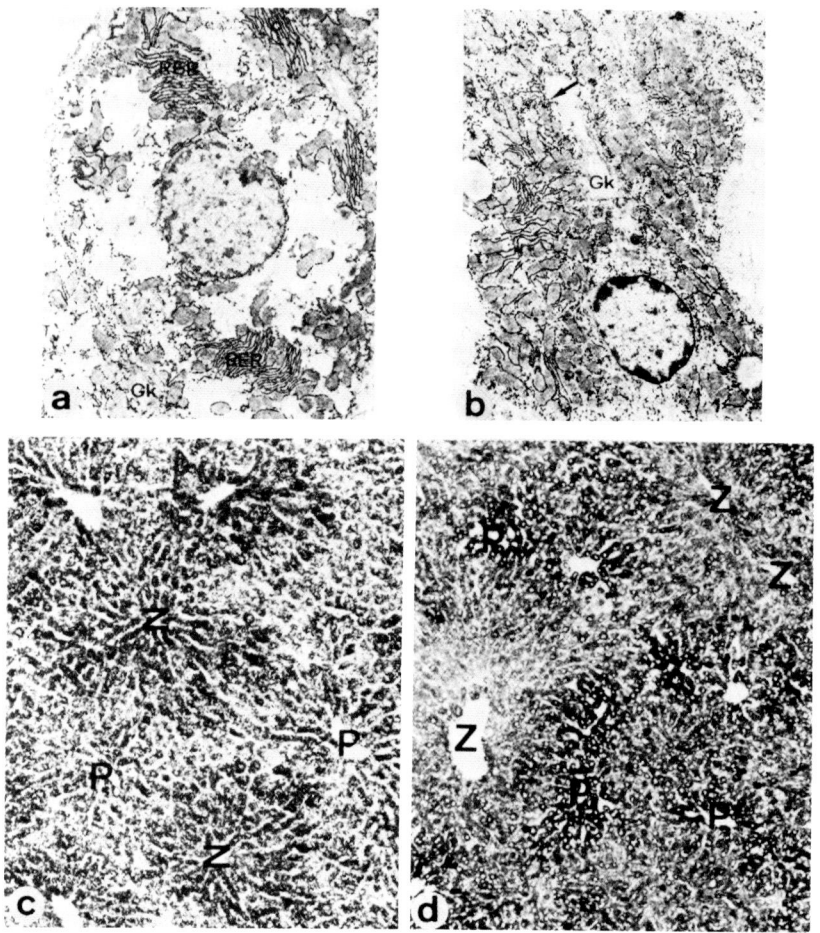

Fig. 4: Electron and light microscopical demonstration of glucose-6-phosphatase (E.C. 3.1.3.9) in the liver of normal ad lib.-fed male rats at the time points of maximal (a,c) resp. minimal (b,d) glycogen content. In the peripheral part of the liver lobule the strongly reacting rER (rough endoplasmic reticulum) is transformed from the stacked (a) into the diffusely spread (b) configuration. Small fields of sER (smooth endoplasmic reticulum) appear in b (arrow). Bile canaliculi (Gk) are tight in a resp. widely expanded in b. P = periportal area, Z = lobular center (from 19).

lishing such time maps as a step towards the understanding of cell work economy and its regulation is obvious. It is also interesting to note that, by using chronobiological protocols in cell cycle research, Burns (4) and Sigdestad (38) came to the conclusion that the estimation of premitotic stage duration may depend on the time of day at which labelled DNA precursors are injected.

As illustrated by fig. 2, the microscopic appearance of salivary gland cells undergoes prominent daytime-dependent changes. These are not simply the result of the passive displacement of cytoplasmic structures by the periodic storage and release of voluminous secretory cell products. Parallel electron microscopic investigations of the same cells (1) have revealed distinctly different ultrastructural configurations, each reflecting a different state of cellular activity with respect to the synthesis, concentration and transport of chemical substances.

Similar to the above glandular cells, the subcellular organization of the multifunctional liver parenchymal cells also displays drastic daytime-associated rearrangements. In the rat liver these changes become most prominent in the perilobular cells at about the timepoints of maximal (at early light) respectively minimal (at early darkness) glycogen content (fig. 3). Along with their changing spatial interrelationship, some cytoplasmic organelles also show circadian variations of their relative volume densities (19,30,39).

The above observations help to underline, first of all, the restricted validity of our understanding of 'normality' with respect to certain cells, inasmuch as we rely only on the results of conventional one-timepoint investigations, which are usually performed at arbitrarily chosen single times during just one and the same half of the 24h-day. Also, that by not being aware of spontaneous circadian stage-specific different trends in the cells' structural-functional organization one may misinterpret the quality and strength of experimental influences. Such error is likely to occur not only when studying short term ("within-working day") effects of experimental treatments. It basically also holds true, e. g., for the afore-mentioned elevation of binucleate hepatocyte indexes due to prolonged conditions of increased metabolic load.

Morphological Aspects of Chronotoxicity

Based on our earlier experience with succinic dehydrogenase rhythmicity in rat liver (23), in a series of basically identical chronotoxicological experiments, we injected groups of rats

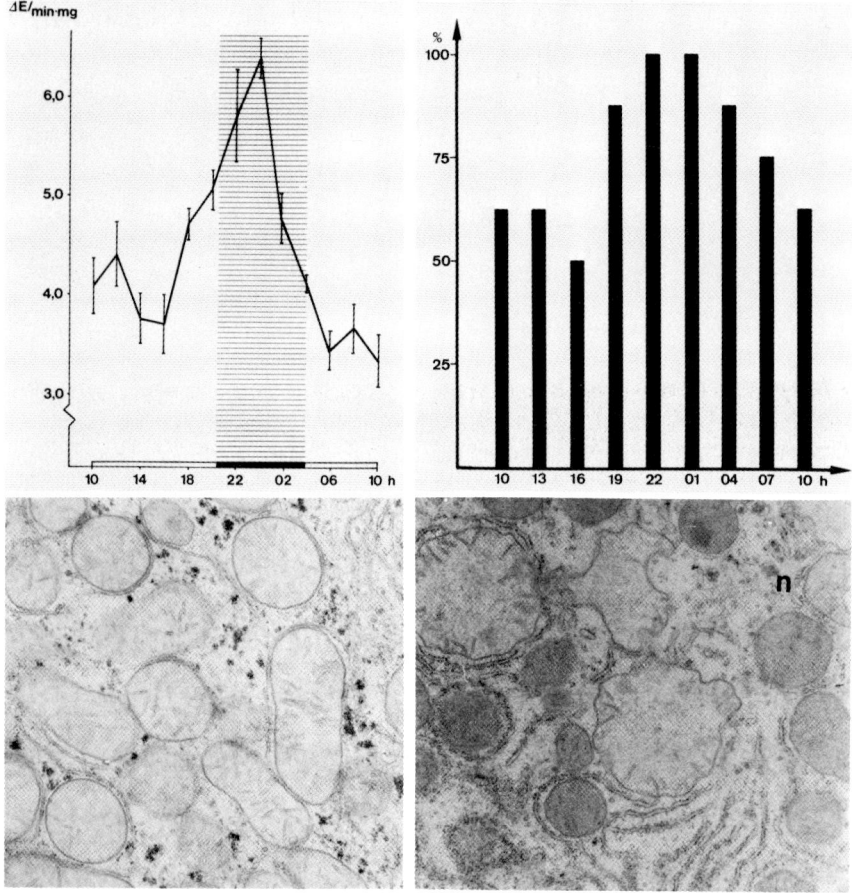

Fig. 5: a = circadian variation of succinate-cy. c reductase (mean values ± s.e., n/timepoint = 4) in isolated liver mitochondria of normal ad lib.-fed rats. b-d = in-vivo chronotoxic effects of antimycin a (1.21 mg/kg b. wt., i.p.) in ad lib.-fed rats; mortality rate (n/timepoint = 8) (b); effect on liver parenchymal cells at 10^{00} (c) resp. 22^{00} (d). Lighting conditions in each investigation were as indicated in a (from 24).

each at a different time of the day but with the same single dose of the respiratory inhibitor antimycin-a. Fig. 5 presents the result of one of these experiments: the chronogram of the mortality rate clearly illustrates the varying responsiveness of the animals to the toxic action of the poison. In good temporal correlation to this, the structural damage of liver cells in antimycin-a-treated animals also varied remarkably as especially visualized by the electron-microscopic appearance of the mitochondria (fig. 5).

From the chronotoxicological studies of Müller (18) and Nair (22), it appeared that the rhythmic variation of hexobarbital oxidase activity in the liver is a rate-limiting factor determining the rhythm of hexobarbital-induced sleep duration in rats. The increase and subsequent decrease of hexobarbital oxidase activity quite well coincides with corresponding changes observed in the relative volume density of the hepatocyte smooth endoplasmic reticulum (sER).

Whereas one might speculate about a direct causal relationship between these quantitative sER variations and hexobarbital chronotoxicity, such a relationship seems not to be responsible for the action of antimycine-a. The degree of hepatocellular damage due to antimycin-a intoxication rather seems to be the combined result of both the circadian differences in mitochondrial sensitivity to respiratory inhibition and the varying antimycin-a-metabolizing potency by sER.

Except for these and few other examples,

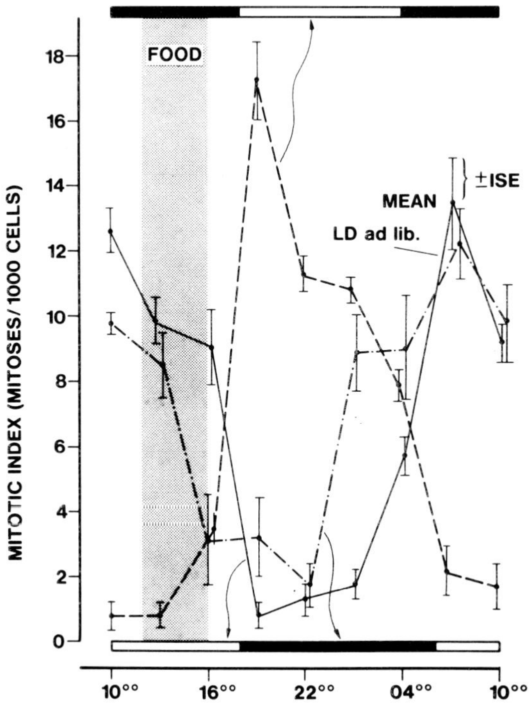

Fig. 6: Environmental light-dark cycle (upper and lower horizontal bars) overrides meal-scheduling (grey) as a synchronizer of mitotic rhythm in rat cornea epithelium (from 28).

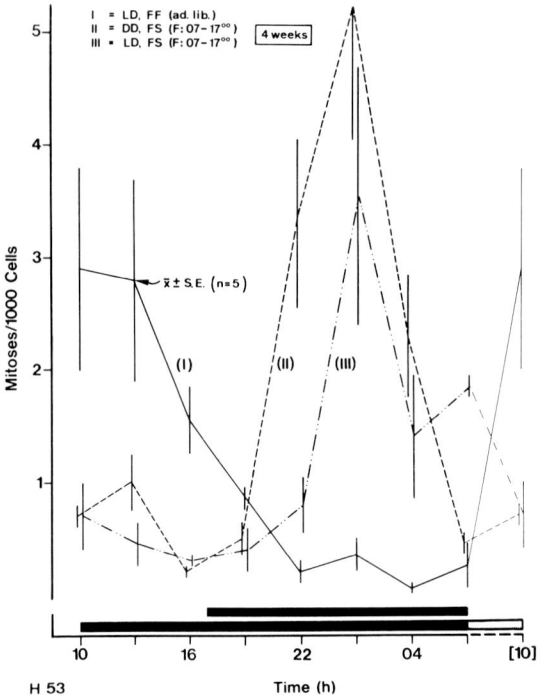

Fig. 4: Meal-scheduling overrides environmental light-dark cycle as a synchronizer of mitotic rhythm in rat liver. LD = light $07^{00}-17^{00}$, darkness $17^{00}-07^{00}$, DD = continuous darkness, FF = ad lib.-feeding, FS = feeding-starvation schedule (from 26).

the vast majority of reports on chronomorphological changes relevant to drug action deal with the evaluation of 24h-rhythmic proliferation indexes and related phenomena in normal and abnormal growth. In biomedical research the ultimate purpose of such evaluations is to enable the optimization of cancer chemotherapy (9,35). As an important part of this, attempts are underway to increase the selective action of cytostatic drugs, e. g., by identifying the times and treating the tumor when it, rather then the normal tissues, is sensitive to the drugs. The chronobiological basis of such manipulatory attempts is the differential influence of periodic environmental inputs on the rhythmic organization of the mammalian system. The graphs in figs. 6 and 7 each clearly exemplify the different synchronizing effects of the light-dark cycle and of meal-timing on the mitotic rhythm of two different tissues in the same organism. Similar differences in synchronizing power may also exist with respect to different functions in a single organ. Based on this, internal phase-disintegration has been observed between a number of biochemically determined metabolic rhythms in rat liver after a sudden phase jump of the environmental light-cycle (25).

CONCLUDING REMARKS

Much to the inconvenience of the experimental investigator, circadian rhythmicity in the living organism complicates the variability derived from other biological factors, like species, strain, substrain, sex and age. These factors, in turn, are known to potentially modify many rhythmic patterns (37). Still more complicating and difficult to analyze are the temporal variations occurring along the circannual time scale ('seasonal rhythms') (14). These variations may occasionally make it difficult to compare experimental data collected at the same "clock hour" of the day but at different times (seasons) of the year.
Against the background of this, one might put the theoretical question, how to escape from time-associated variability in biological experimentation. The answer can not be by eliminating the more commonly known environmental cycles of light-and-darkness, ambient temperature, air humidity or nutritional input. Indeed, experimental protocols of this kind have been the basis of a multitude of so-called synchronizer studies. One interesting finding, among many others, was the persistence of the histochemically demonstrable, 24-hour periodicity of some enzyme activities (succinic dehydrogenase, acid phosphatase) in the intestinal mucosa of hibernating dormice (Glis glis), even after three months without food (32).

REFERENCES

1. Albegger K, Müller O (1973). Zur Circadianstruktur der Glandula submandibularis. Arch klin exp Ohr Nas Kehlk Heilk 205:122.
2. Barnum CP, Jardetzky CD, Halberg F (1958). Time relations among metabolic and morphologic 24-hour changes in mouse liver. Am J Physiol 195:301.
3. Boehm N, Moser B (1976). Reversible Hyperplasie und Hypertrophie der Mäuseleber unter funktioneller Belastung mit Phenobarbital. Beitr Path 157:283.
4. Burns ER, Scheving LE (1975). Circadian influence on the wave form of the frequency of labeled mitoses in mouse corneal epithelium. Cell Tissue Kinet 8:61.
5. Dvorak M (1974). Origin and development of lysosomes and peroxisomes. In Dvorak M (ed): "Biogenesis of Cell Organelles." Acta Fac Med Univ Brunesis.
6. Echave Llanos JM, Aloisso MD, Souto M, Balduzzi R, Surur

JM (1971). Circadian variations of DNA synthesis, mitotic activity and cell size of hepatocyte population in young immature male mouse growing liver. Virch Archiv (Cell Path) 8:309.
7. Forsgren E (1928). Mikroskopische Untersuchungen über die Gallenbildung in den Leberzellen. Z Zellforsch 6:647.
8. Groh V, Mayersbach Hv (1979). Histochemical approach to the question of the heterogeneity of lysosomes. Chronobiologia 6:102.
9. Halberg F, Haus E, Cardoso SS, Scheving LE, Kühl JFW, Shiotsuka R, Rosene G, Pauly JE, Runge W, Spalding JE, Lee JK, Good RA (1973). Toward a chronotherapy of neoplasia: tolerance of treatment depends upon host rhythms. Experientia 29:909.
10. Hardonk MJ, Koudstaal J (1976). Enzyme histochemistry as a link between biochemistry and morphology. Progr Histochem Cytochem 8:68
11. Holmgren H (1936). Studien über 24-Stunden rhythmische Variationen des Darm-, Lungen- und Leberfettes. Dissertation Helsingfors.
12. Klaushofer K, Mayersbach Hv (1977). 5-Nucleotidase der Rattenleber. Spezifitätsprobleme und rhythmische Veränderungen. Acta histochem Suppl XVIII:159.
13. Malis F, Lojda Z, Fric P, Slaby J (1977). Zirkadianrhythmus einiger Verdauungsenzyme des Dünndarms und des Pankreas bei Meerschweinchen. III Bilaterales Symp CSSR-DDR Fortschritte der Gastroenterologie. Karlovy Vary, p 182.
14. Mayersbach Hv (1967). Seasonal influences on biological rhythms of standardized laboratory animals. In Mayersbach Hv (ed): The Cellular Aspects of Biorhythms," Heidelberg: Springer, p 87.
15. Mayersbach Hv (1980). Rhythms at morphological levels. In Scheving LE, Halberg F (eds): "Chronobiology: Principles and Applications to Shifts in Schedules." Alphen aan den Rijn: Sijthoff & Noordhoff, p 95.
16. Mayersbach Hv, Leske R (1963). Physiologische Schwankungen des Glykogen-Tagesrhythmus. Acta morph Acad Sci hung XII: 33.
17. Mayersbach Hv, Philippens K, Yap P (1964). Die Einflüsse biologischer Tagesschwankungen auf fermenthistochemische Untersuchungen. II Internat Congr Histochem Cytochem 1964. Berlin-Göttingen-Heidelberg: Springer, p 139.
18. Müller O (1974). Circadian rhythmicity in response to barbiturates. In Scheving LE, Halberg F, Pauly JE (eds): "Chronobiology," Tokyo: Igaku Shoin Ltd, p 187.
19. Müller O (1977). Der circadiane Strukturwandel der Leber

und der Speicheldrüsen. Nova Acta Leopoldina 225:131.
20. Müller O, Preuss D (1976). Circadiane Histochemie der Glycogen-Synthetase und -Phosphorylase in der Leber der Ma s. XVII Symp Ges Histochemie Bozen (1974). Acta histochem Suppl XVI:145.
21. Münzer FT (1924). Experimentelle Studien über die Zweikernigkeit der Leberzellen. Arch Mikr Anat 104:138.
22. Nair V (1974). Circadian rhythm in drug action: A pharmacologic, biochemical, and electronmicroscopic study. In Scheving LE, Halberg F, Pauly JE (eds): "Chronobiology," Tokyo: Igaku Shoin Ltd, p 182.
23. Philippens KMH (1971). Vergleichende Untersuchungen über biochemische Aktivitätsbestimmungen an Mitochondrien und histochemischem Reaktionsausfall. Acta histochem Suppl X:323.
24. Philippens KMH (1974). Circadian variations in rat liver mitochondrial activity. In Scheving LE, Halberg F, Pauly JE (eds): "Chronobiology," Tokyo: Igaku Shoin Ltd, p 23.
25. Philippens KMH (1976). The manipulation of circadian rhythms. Arch Toxicol 36:277.
26. Philippens KMH (1980). Synchronization of rhythms to meal-timing. In: Scheving LE, Halberg F (eds): "Chronobiology: Principles and Applications to Shifts in Schedules," Alphen aan den Rijn: Sijthoff & Noordhoff, p 403.
27. Philippens KMH, Abicht J (1975). Tagesrhythmik des Nukleinsäurestoffwechsels. Nova Acta Leopoldina 225:143.
28. Philippens KMH, Mayersbach Hv, Scheving LE (1977). Effects of the scheduling of meal-feeding at different phases of the circadian system in rats. J Nutrition 107:176.
29. Polak JM, Pearse AGE, Mourik Mv, Mayersbach Hv (1975). Circadian rhythm of the endocrine pancreas. A quantitative biochemical and immunocytochemical study. Acta Hepato-Gastroenterol 22:118.
30. Rohr HP, Hundstad AC, Bianchi L, Eckert H (1970). Morphometrisch-ultrastrukturelle Untersuchungen über die durch die Tageszeit induzierten Veränderungen der Rattenleberparenchymzelle. Acta anat 76:102.
31. Ruby JR, Scheving LE, Gray SB, White K (1973). Circadian rhythm of nuclear DNA in adult rat liver. Exp Cell Res 76:136.
32. Sauerbier I (1977). Die Circadianrhythmik während des Winterschlafes. Biochemisch und histochemisch-morphologisch faßbare Veränderungen im Verdauungstrakt von Glis glis. Nova Acta Leopoldina 225:163.
33. Sauerbier I (1978). Circadianrhythmisch-histochemische Untersuchungen am Dünndarmepithel von Maus und Huhn. Acta

histochem 63:214.
34. Scheving LE (1973). Cellular mechanisms involving biorhythms with emphasis on those rhythms associated with the S and M stages of the cell cycle. Internat J Chronobiology 1:269.
35. Scheving LE, Burns ER, Pauly JE, Halberg F, Haus E (1977). Survival and cure of leukemic mice after circadian optimization of treatment with cyclophosphamide and 1-ß-D-Arabinofuranosylcytosine. Cancer Research 37:3648.
36. Scheving LE, Burns ER, Pauly JE, Tsai TH (1978). Circadian variation in cell division of the mouse alimentary tract, bone marrow and corneal epithelium. Anat Rec 191:479.
37. Scheving LE, Halberg F, Pauly JE (1974). "Chronobiology." Tokyo: Igaku Shoin Ltd.
38. Sigdestad CP, Bauman J, Lesher SW (1969). Diurnal fluctuation in the number of cells in mitosis and DNA synthesis in the jejunum of the mouse. Exptl Cell Res 58:159.
39. Uchiyama Y, Mayersbach Hv (1979). Study on the rhythmic changes of lysosomal enzyme activities in rat liver parenchymal cells using ultracytochemistry. Chronobiologica 6:166.
40. Vonnahme FJ (1974). Circadian variation in cell size and mitotic index in tissues having a relatively low proliferation rate in both normal and hypophysectomized rats. Internat J Chronobiology 2:297.
41. Zaviacic M, Brozman M (1978). Circadian rhythms of oxydoreductases in the rat gastric mucosa. Histochemical study. Acta histochem 62:155.

CIRCADIAN RHYTHMS IN CELL PROLIFERATION: THEIR IMPORTANCE WHEN INVESTIGATING THE BASIC MECHANISM OF NORMAL VERSUS ABNORMAL GROWTH

Lawrence E. Scheving

Department of Anatomy

University of Arkansas for Medical Sciences,
Little Rock, Arkansas 72205

HISTORY OF RESEARCH ON CELL-DIVISION RHYTHMS

Plants

One of the first reports in the literature suggesting a circadian rhythm in cell division in plants was by Killicott (1904). He reported that the maximum mitotic index in the Allium root tip occurred at 2300 and the minimum at 0700. Karsten (1918) reported a similar phenomenon for the mitotic index of Spirogyra, with highest activity during mid-day, and for Zea mais with the highest mitotic index occurring during the night. Stalfelt (1921) reported that the highest mitotic index in Pisum sedativum occurred at mid-day and the minimum during the morning. With these and several other early studies, the botanists pioneered in exploring rhythms in cell division.

Animals

Fortuyn-van-Leyden (1917, 1926) was the first to report on mitotic-index rhythms in animals, specifically on rhythmic variations in several tissues of the cat. Most of these studies used once-a-day or once-a-night sampling and small numbers of animals; consequently her findings in some cases were contradictory from study to study. One only has to examine the data she collected, however, to realize that Fortuyn-van-Leyden was correct in her suggestion that cell division in the cat was periodic and, therefore, in this respect was not unlike cell division in plants.

In the 1930's and 1940's, there was renewed interest in this phenomenon, and several papers appeared. Carleton (1934), Ortiz-Picón (1934) and Cooper and Franklin (1940) all reported a rhythm in the epidermis of the mouse. Although these studies were very limited as to sampling intervals and numbers of animals used, generally mitotic activity was found to be highest about mid-day and lowest at night. The work of Ortiz-Picón (1934) was, up to that time, the most extensive in that he sampled at four time points, 1200, 1900, 2000 and 2400, and used six mice per time point. In reality, however, the above represents only three circadian stages since the 1900 and 2000 time points are so close together.

More extensive, and thus more reliable, investigations were made on rhythms in cell division in the rat by Blumenfeld (1939; 1943). He found that the maximum mitotic index of the abdominal epidermis occurred between 0800 and 1000, and the minimum occurred between 2000 and 2400. In these same rats, he also determined the mitotic index in the submandibular gland and kidney. Although the times of peak mitotic index were different for these three tissues, it was of interest to find that the time of the lowest mitotic index was identical for all. Blumenfeld discussed at length the differences in the timing of the rhythms, but did not discuss the similarity of phasing of the lowest values. We have found more often than not that the timing of the low point of a rhythm is far less variable than is that of the peak of the rhythm.

Man

In 1938, Cooper and Schiff analyzed mitosis in the prepuce of 8-day-old infants, and concluded that the highest mitotic index occurred at 0900 and the lowest at 2200; only 13 specimens were used to evaluate the 24-hour span, and no sampling was performed between 1245 and 0730. A similar study followed in 1939, when Broders and Dublin examined 14 foreskins and concluded that the highest mitotic activity occurred at night. No sampling was done between 1200 and 0730.

It was against the above background that my interest in research in rhythmicity and cell division arose. I was aware of the fact that anatomists and pathologists were bewildered by the sparse number of mitotic figures that frequently were found in histological preparations. This especially was puzzling for the epidermis which we knew was continually being shed and renewed. The question often asked was whether the few mitotic figures frequently encountered in biopsy specimens (more often than not obtained in the morning) could account for the renewal of all of the skin that was known to be lost during the course of a single day? Some said "no"; and several investigators including Frieboes (1920), Bostroem (1928), Cameron (1936), Levander (1950) and others had postulated that epidermal cells may have a mesodermal origin. It was in 1949 that I read a report by Andrew and Andrew introducing the hypothesis that in the human epidermis migrating lymphocytes became transformed into "clear cells" which later differentiated into epidermal cells. Andrew (a histologist) also believed that this was the reason why normally one did not find many mitotic figures in skin specimens found in the histological or pathological slide sets used for teaching students.

This explanation was difficult to accept; and it seemed more logical to me, in light of the reports on rhythmicity, that the paucity of mitotic figures might well be explained by the fact that mitosis occurs in daily cycles. Therefore, a study was designed to test this theory and to evaluate any evidence of lymphocyte transformation into epithelial cells.

In retrospect, as will be seen, it was unfortunate that I selected the infant prepuce to investigate for circadian rhythmicity. This selection was prompted principally by the availability of this tissue, and also because I felt that the authors of the two earlier studies on infant prepuce had based their conclusions on too few specimens. With the aid of a colleague who had been my undergraduate college roommate (Mr. Joseph P. Young), I began collecting foreskins. Mr. Young was a medical representative with Lederle Laboratories and he called periodically on several hospitals in the Chicago area. Mr. Young was an able salesman, both for his company and for his ability to sell physicians on the idea of performing circumcisions on the newborn at different hours of the day and night. Thus, we were able to collect over 150 foreskins representing circumcisions performed in the course of about one and a half years, all along the 24-hour time scale.

The end result of this large study was that no evidence of a circadian rhythm in the mitotic index was found; at least it was not evident from plotting the data as a chronogram or after analyzing with conventional statistics. Moreover, we found no evidence of lymphocyte transformation (Scheving, 1959). What appeared in the data was the suggestion of an ultradian rhythm as there were multiple peaks and troughs in the mitotic index along the 24-hour time span. In short, we did not confirm the findings of Cooper and Schiff (1938) or Broders and Dublin (1939). This was rather discouraging; and what seemed to be "negative" data could not be published easily since they tended to refute the two earlier published works that seemingly had confirmed one another, albeit with very scanty data. Therefore, our data were put in a desk drawer where they remained for a couple of years. Later I moved to another city, and the foreskin slides and data sheets were stored in a box in a "crawl space" beneath my new house. Shortly thereafter this space was inundated with several feet of flood water; and the slides, labels and data sheets became wet and impossible to read, and everything was thrown out. I hope that in the future someone will repeat this study because we now know that the circadian rhythm in the mitotic index of corneal epithelium in the rodent does not manifest itself until about the 15th day of postnatal life (Goodrum et al., 1974).

During the early 1950's the hypothesis of Andrew was becoming more and more talked about, and because of this, I decided to study my own skin. With the aid of my wife, I removed skin samples from my own shoulders using a punch biopsy, obtaining samples of skin during the morning and after midnight on several occasions. There clearly was a difference in the mitotic indices between the two time points, with the highest mitotic activity shortly after midnight. My younger brother volunteered skin samples at similar times on one of his visits to our home, and they served to confirm this preliminary observation.

A more extensive investigation that grew out of these preliminary studies (Scheving and Gatz, 1955; Scheving, 1959) found that the mitotic-index rhythm in the epidermis of the adult man clearly was circadian with the great bulk of cell division taking place after midnight. These data are illustrated in Figure 3 of the first paper of this symposium by Pauly, along with data confirming our findings by Fisher (1968). It is of interest that this finding was again confirmed recently, over 20 years later (Zugula Mally et al., 1979).

In addition to the skin and the extensive documentation of the remarkable circadian variation found in human circulating blood cells, there are, to the best of my knowledge, only two other reports in the literature showing a mitotic-index rhythm in man and that is in the bone marrow (Killman et al., 1962; Mauer, 1965). The data from these two studies are illustrated in Figure 1. Although data of this kind are sparse, it must be kept in mind that serial bone-marrow biopsies are very difficult to obtain from healthy subjects.

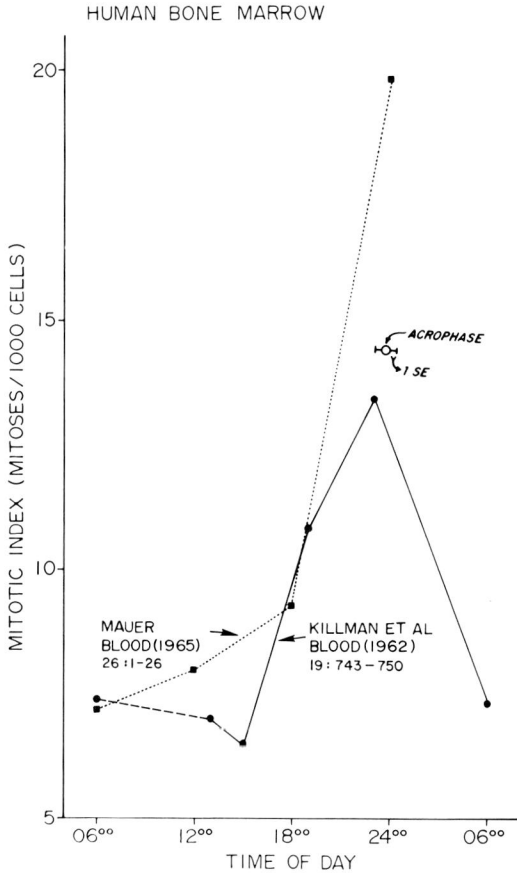

Fig. 1. The mitotic-index rhythm in the bone marrow of a group of young men (dotted line) and the mitotic rhythm in the bone marrow obtained by repeated biopsy from a single individual along a 24-hour span. For details, see Mauer (1965) and Killman et al. (1962).

It is emphasized that man is a very rhythmic creature. Not only are there rhythms in cell division, and circulating levels of blood cells, but also man shows a remarkably rhythmic time structure for vital signs, performance and in a host of variables in blood and urine. Figure 2 is an acrophase map showing a large number of rhythmic variables, all of which were studied on the same group of men. Since this acrophase map was first drawn, we and others have found over forty additional variables in blood, urine and saliva which also undergo circadian fluctuation (Kanabrocki et al., 1974; Scheving et al., 1977a).

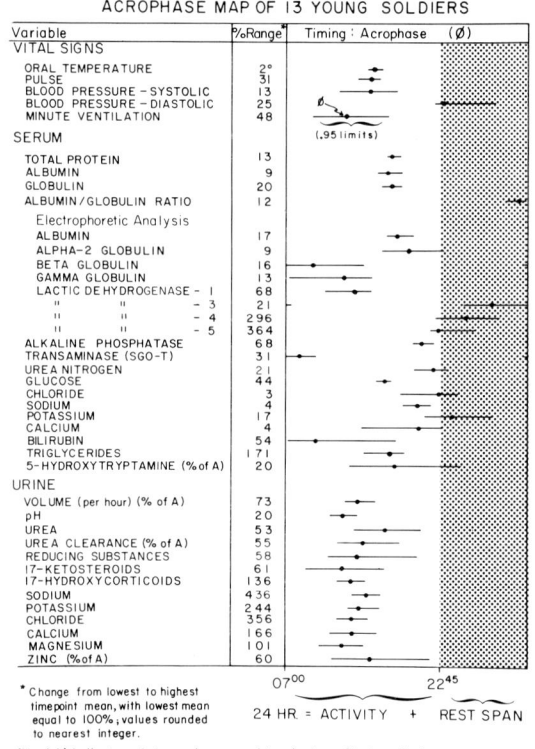

Fig. 2. The acrophase (∅) for each variable approximates the peak or crest of the circadian variation. The percent range of change reflects the change from the lowest to the highest time-point means, with the lowest mean equal to 100% (Kanabrocki et al., 1974; Scheving et al., 1977a).

More Recent Studies on Cell Division in Animals

In the 1950's there was an increase in interest in the circadian rhythm in cell division in animals. In a study by Barnum, Jardetzky and Halberg (1957) on regenerating liver, radioactive isotopes were introduced so that not only the mitotic index but also the S phase, could be evaluated. The S phase was also shown to undergo prominent circadian variation.

Scheving and Pauly (1960) with sampling at 1-hr intervals found the peak mitotic index in the epidermis of the pinna of the rat to occur somewhat earlier than that reported by Blumenfeld (1939). We later confirmed and extended our own findings (Scheving and Pauly, 1977) (Fig. 3). It should be noted, however, that we do not know anything about the light-dark cycle to which Blumenfeld's animals were subjected. Tvermyr (1969) in an extensive study on hairless mice also reported a rhythm in the mitotic index of the epidermis with a phasing similar to that we have reported. In addition to the epidermis, we also carried out extensive studies on the mitotic indices of the cornea, tongue, ameloblasts and duodenum of the rat, all of which demonstrated overt circadian rhythms (Scheving and Pauly, 1967; Gasser et al., 1972a,b; Scheving and Pauly, 1977).

There were, however, during the 1960's a series of studies which used colchicine to collect mitotic figures over four different 6-hr spans along the 24-hr time scale. This technique unfortunately produced negative results for detecting rhythms in a number of tissues, including the intestinal tract, uterus and cornea. The major contributors reporting a lack of rhythmicity were Bertalanffy and his colleagues. These studies and other studies such as that by Halberg et al. (1959) have been reviewed by us (Scheving and Pauly, 1973; Scheving et al., 1981a). The collection of data and the evaluation of rhythmic events in cell division continues to the present time; and we can now conclude that all normal tissues in adult mammals examined to date undergo circadian variation.

Such fluctuations are evident whether we count mitotic figures or measure the incorporation of tritiated thymidine, (^{3}H)TdR, into DNA or utilize the newer cytoflurometric technique of analyzing cell proliferation (Lareum and Aardal, and Rubin; this symposium). The amplitudes of rhythms clearly vary from tissue to tissue and within different

RHYTHMOMETRIC SUMMARY

	P	Mesor, M M ± SE	Amplitude, A A ± SE	SE/A	Acrophase, Ø (95 confidence limits)
A	.20	4.4 ± .2	6 ± .3	.538	−152°
B	.01	3.1 ± .2	1.2 ± .3	.287	−82° (−50°, −115°)
C	.001	7.6 ± .5	3.3 ± .7	.214	−51° (−27°, −75°)

Acrophase reference = local midnight

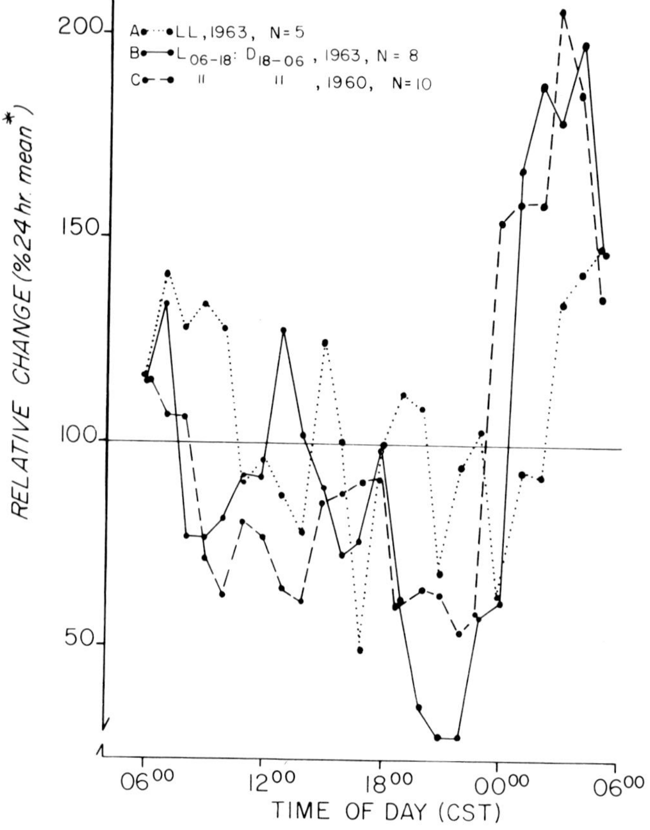
A●···●LL, 1963, N = 5
B●—●L06-18: D18-06, 1963, N = 8
C●- -● " " , 1960, N = 10
RELATIVE CHANGE (% 24 hr. mean*)
200
150
100
50
06 00
12 00
18 00
00 00
06 00
TIME OF DAY (CST)

*Mitoses / 1000 cells (2500–3000 cells / specimen)

Fig. 3. Mitotic index rhythms in the pinna epidermis of rats. A (dotted line): data from rats maintained in continuous illumination (LL, 600 lux). These data were not significant as indicated by the high "noise-to-signal ratio" (SE/A) which is an earlier method of estimating the significance of the data using the cosinor technique (Pauly, 1981; Halberg et al., 1967, 1972). B and C (solid and dashed lines): data from animals synchronized to a light-dark cycle (LD 12:12); the phasings of these rhythms are very similar and the confidence limits of the two acrophases do overlap (Scheving and Pauly, 1960, 1977).

regions of an organ or tissue, however, some having very high and others rather low ranges of change along the 24-hr time scale. Such variation is evident in Figure 4 which illustrates rhythms in the incorporation of (^{3}H)TdR into DNA of the ovary and in the mitotic index of the cornea from the same animals (Scheving et al., 1981a). These represent extremes of high-and low-amplitude rhythms. Figure 5 is another example of a high-amplitude rhythm in mitotic index

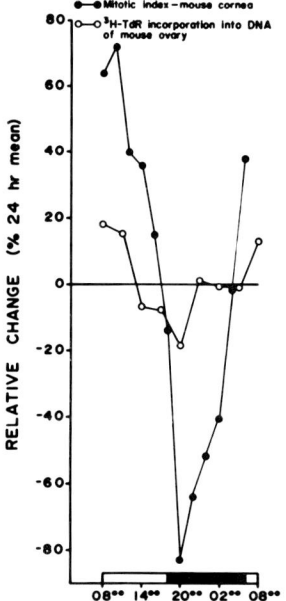

Fig. 4. Chronograms illustrating the mitotic-index rhythm in the corneal epithelium of mice and the rhythm in the incorporation of (^{3}H) TdR into DNA of the ovary of the same mice (Scheving et al., 1981a).

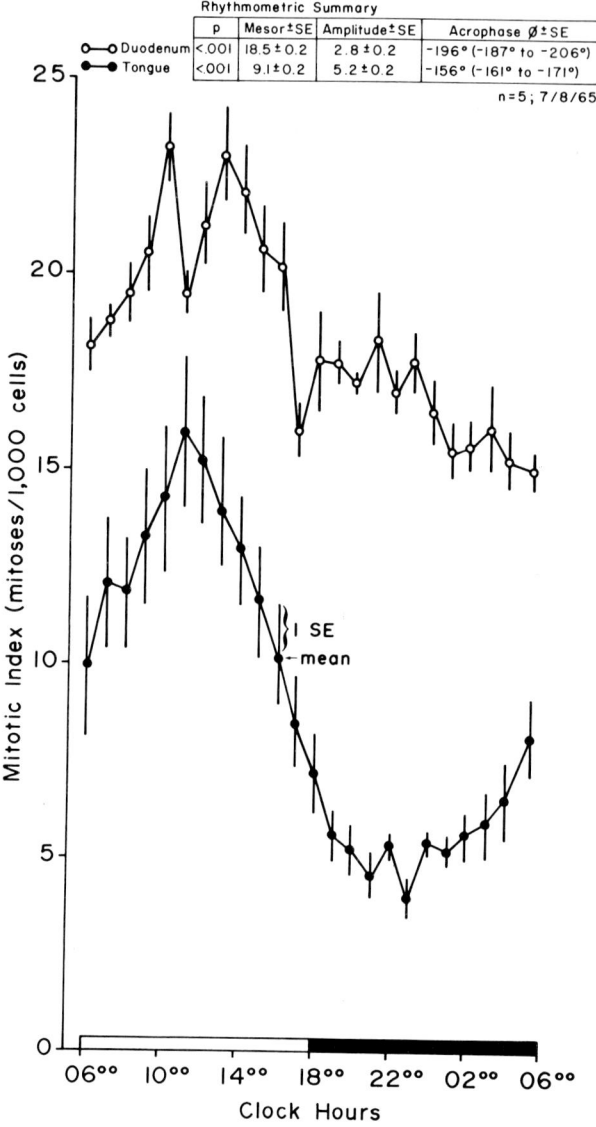

Fig. 5. Chronograms illustrating the mitotic indices in the duodenum and tongue of the same animals (duodenum data from Scheving and Pauly, 1973; tongue data from Gasser et al., 1972a).

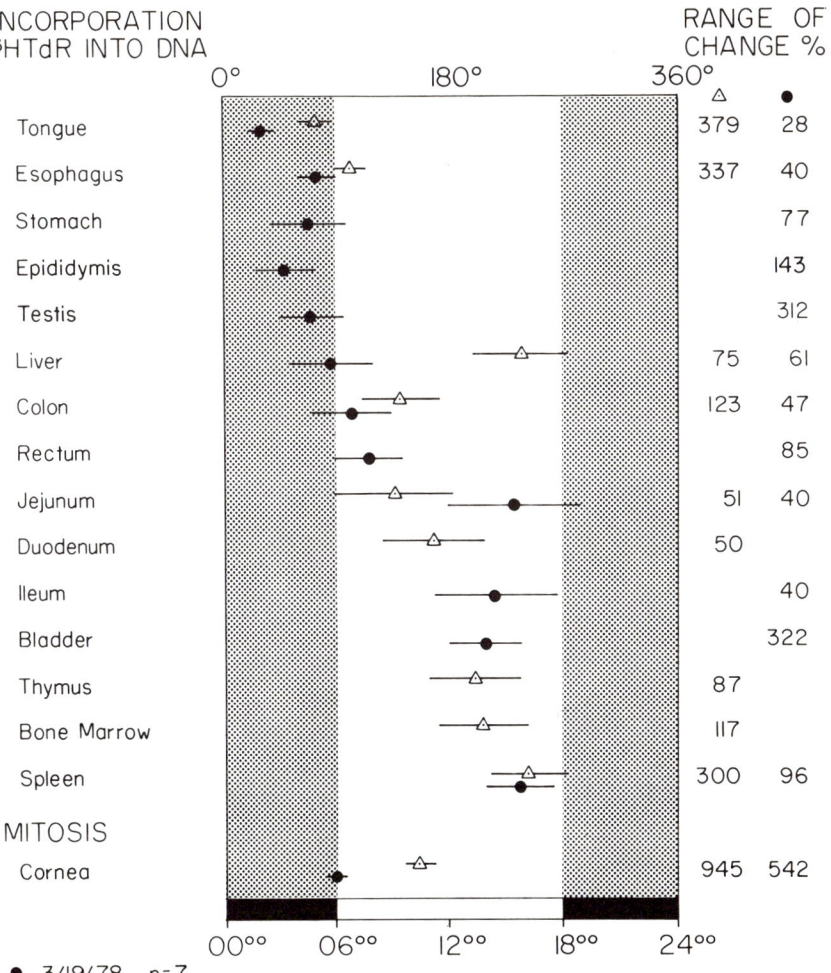

Fig. 6. An acrophase map, showing the approximate peak of the circadian cycle in either the incorporation of $(^3H)TdR$ into DNA or in the mitotic index of the corneal epithelium of mice. The acrophase ($\emptyset$) is represented by a dot or triangle, and the bars extending out from the dots or triangles represent the confidence limits. The shaded portion represents the 12-hour dark phase of the environmental light-dark cycle. The percent range of change reflects the change from the lowest to highest time-point means, with the lowest mean equal to 100%; values rounded to nearest integer (Scheving et al., 1981a).

of the tongue, and a low-amplitude rhythm in mitotic index of duodenal epithelium of the same rats. Most of the digestive tract both proximally and distally to the duodenum has a higher amplitude rhythm in the mitotic index; in our opinion the higher amplitude type of rhythm is the most common. Another characteristic of cell division in the rodent is the finding that in some tissues, although rhythmic, cell divisions is always occurring at a comparatively high level; the duodenum is an excellent example of this type of tissue. In other tissues, such as the cornea (Fig. 4), both DNA synthesis and the mitotic index drop to practically zero at certain circadian stages. Figure 6 is an acrophase map showing (^{3}H) TdR incorporation into DNA in 15 different tissues of the mouse and the mitotic index of corneal epithelium.

SYNCHRONIZATION OF RHYTHMS

To a Light-Dark Cycle

All rhythms in cell division are synchronized to the light-dark cycle if the animals are fed ad libitum. Figure 7 illustrates a mitotic-index rhythm in the corneal epithelium which can be inverted within 7 days or less simply by inverting the ambient light-dark cycle 180°. The rhythms in cell division in other tissues may take longer to invert; for example, the rhythm in the incorporation of (^{3}H)TdR into DNA in the spleen takes 3 weeks to invert (unpublished results). For a more extensive treatment of synchronization of rhythms to the light-dark cycle see Scheving and Pauly (1973), and Scheving et al. (1981a).

To Meal Timing

Some rhythms in cell-division such as those in the bone marrow, spleen and intestinal tract can be manipulated by restricting feeding to a particular time of the day. For these tissues meal-timing is capable of partially, or even completely, over-riding the effect of the light-dark cycle (Scheving et al., 1976). The same has been reported for the circulating eosinophils in these same mice (Pauly et al., 1975). Other tissues, such as the corneal epithelium (Scheving et al., 1974a), cannot be easily manipulated by restricted feeding, indicating that the light-dark cycle is a far more dominant synchronizer for the cell-division rhythm in these tissue than it is for the gut, bone marrow, spleen,

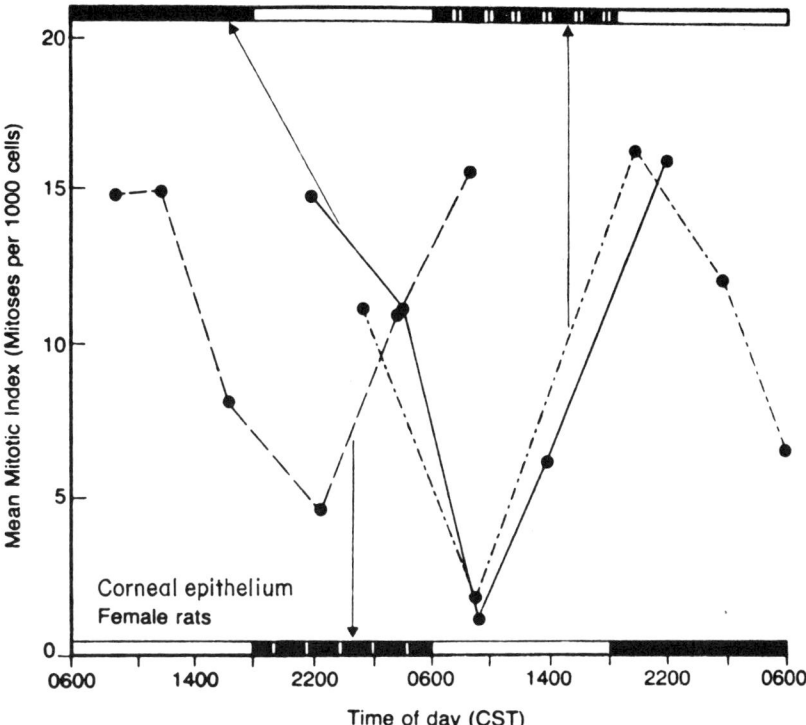

Fig. 7. The curve on the left (dashed line) illustrates a typical mitotic-index rhythm in rats standardized to 12 hours of light alternating with 12 hours of darkness shown on the bottom abscissa. The dotted-dashed curve to the right represents the inverted pattern seven days subsequent to inverting the light-dark cycle (see horizontal axis on top of the figure). The solid line accompanying it represents another study conducted 12 days later showing that the rhythm was "locked in" to the light-dark cycle. In all cases N=5 (Scheving and Pauly, 1973; Scheving et al., 1981b).

or eosinophils. Philippens et al. (1977) confirmed this finding for the cornea of the rat as well as carrying out additional studies on the effect of meal timing on a host of biochemical variables. Figure 8 illustrates the lack of much response in the mitotic index in corneal epithelium to restricted feeding schedules (Scheving et al., 1974a); Figure 9 illustrates a dramatic response in circulating eosinophils to the same treatment in the same mice (Pauly et al., 1975).

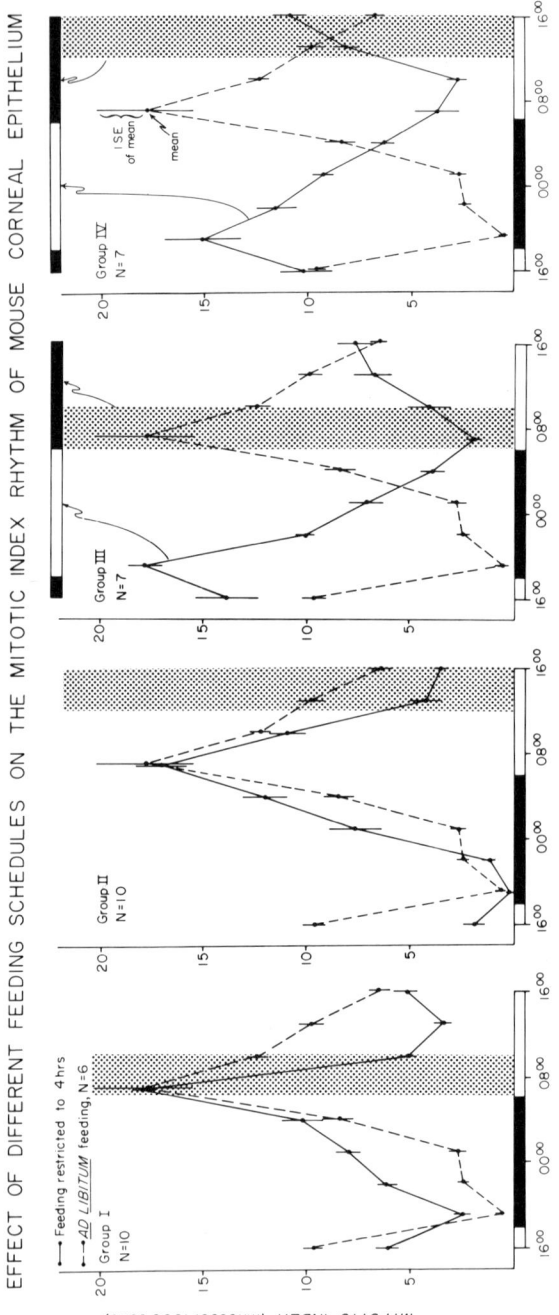

Fig. 8. Effect of feeding schedules on the mitotic index rhythm in the rat corneal epithelium. The light-dark cycle to which the animals were subjected is depicted on the horizontal axis of each chronogram. Within each group of mice, controls were fed ad lib and experimental mice were restricted in their access to food each day to 12 hours during the first week, to 8 hours the second week, to 6 hours the third week, and to 4 hours during the fourth and fifth weeks (shaded areas). Note that the mitotic-index rhythms remained remarkably synchronized to the light-dark cycle. In all instances, time was local clock time (CST). For details of this study, and for the data shown in Figure 9, see Scheving et al. (1974a).

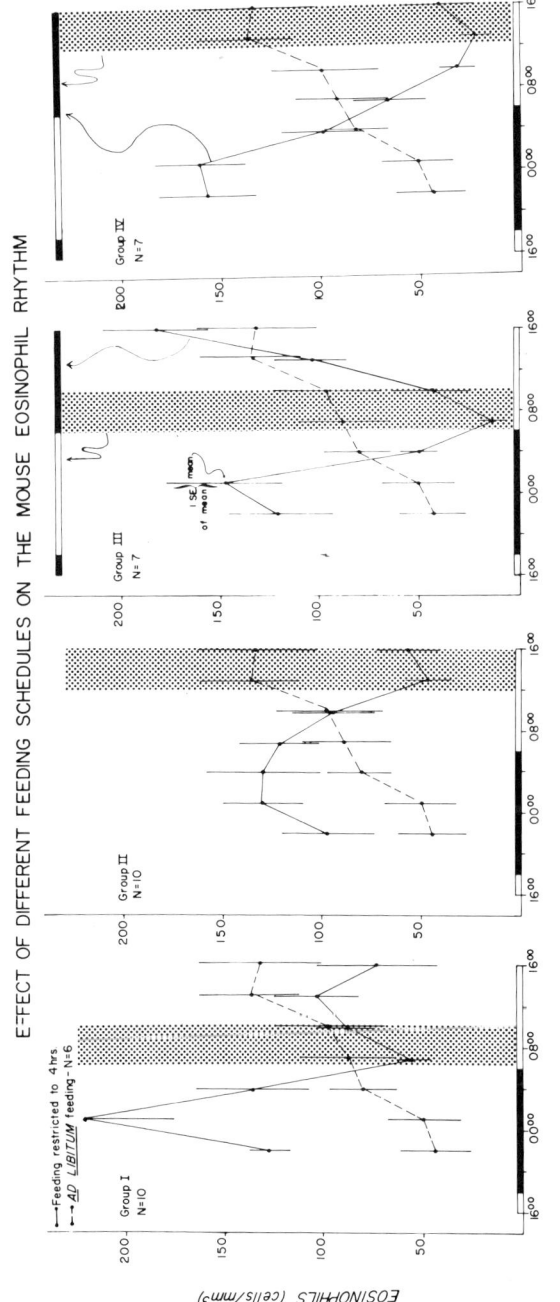

Fig. 9. The effect of restricted feeding schedules on eosinophil levels in mice in the same animals described in Figure 8 (the data were published earlier by Pauly et al., 1975). Unlike the case for the mitotic index of the corneal epithelium, the feeding schedule strongly synchronized the eosinophil rhythm.

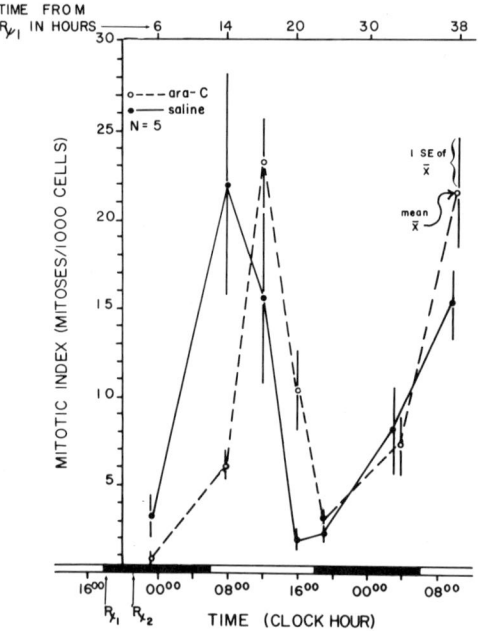

Fig. 10. Phase shift in the circadian rhythm of mitosis in mouse corneal epithelium induced by arabinosyl cytosine (ara-C). For explanation see text as well as Scheving and Pauly (1973).

The data obtained from the extensive studies performed by ourselves and many others support the following conclusions, not only for rhythms in cell division but for rhythmic variables in general: (1) The synchronizing effects of light and food cannot be generalized with respect to all variables within the organism's circadian system. (2) Food is managed differently when given at different stages of the circadian system; it simply is not enough to give the meal at a single stage of the organism's circadian system, as frequently has been done, and assume that a particular response would be the same if the meal were presented at another stage. (3) Restricted feeding may prove to be a very useful tool to facilitate investigation of control mechanisms of cell division or to improve cancer chemotherapy. We have postulated that because of the ability of restricted feeding schedules to synchronize, at least partially, cell proliferation in bone marrow, spleen and intestinal tract, this might be used to maximize chemotherapy in the clinics (Scheving et al., 1976). Although this hypothesis has been supported by studies on experimental chemotherapy in the rodent (Nelson et al., 1974), it has not been tested in the clinics.

PHASE SHIFTING OF RHYTHMS

It has been shown repeatedly that drugs can alter the rhythm of the mitotic index (Scheving and Pauly, 1971, 1973; Burns and Scheving, 1973). Figure 10 illustrates the results of an investigation performed on adult mice standardized 7 days prior to the beginning of the study to artificial light from 0600 to 1800, alternating with darkness from 1800 to 0600, with food and water available ad libitum. Forty BD_2F_1 mice were injected intraperitoneally with 2.6 mg of arabinosyl cytosine (ara-C) and 0.2 ml of saline. Two hours after the second injection, subgroups of five mice from both the control and the experimental groups were sacrificed at intervals indicated in Figure 10 for 36 hours. The mitotic rate of the corneal epithelium in the saline-injected animals showed the expected circadian rhythm and the phasing corresponded to that previously found in similarly standardized rats. In the ara-C-injected animals, however, a phase delay of about 4 hours was apparent on the first day after administration of the drug; on the second day after treatment, the usual phase relationship was re-established between the mitotic rhythm and the synchronizer cycle of alternating light and darkness (Scheving and Pauly, 1973).

The rationale for performing the experiment dealing with the effect of ara-C on the mitotic rate in the corneal epithelium stemmed from a series of earlier experiments (Scheving and Pauly, 1971). It had been found that when ara-C was injected during the circadian-system stage at which the rate of corneal mitosis is usually highest, no statistically significant decrease in mitotic count could be detected 12 or 24 hours later. The mitotic peak occurred 24 hours later at the expected time. When the same dose of ara-C was injected at the circadian system stage when the corneal mitotic rate usually is low, however, there usually was a significant decrease in mitotic rate the next day 12 hours later at the time of the expected circadian high. The question asked was whether this apparent decrease was the result of a phase shift in the rhythm, or whether it actually represented a depression in mitotic rate? From examination of Figure 10, it appears that the results obtained were due to a phase delay of about 4 hours. Single time-point sampling could be quite misleading in this situation because it might suggest that a depression of mitosis had taken place when actually there was a phase delay in the rhythm. We recognize, of course, that ara-C can depress mitosis in corneal epithelium, if given at sufficiently high doses, for a longer duration of time, or perhaps even at a different stage of the circadian system. The data in Figure 10 do, however, point out clearly the complexity of evaluating the effect of a drug on mitotic activity; they demonstrate the pitfalls awaiting those who ignore the organism's time structure in the circadian and other frequency ranges.

MECHANISMS OF RHYTHMS IN CELL PROLIFERATION

In spite of much research into the mechanism of cell division and its associated rhythmicity it is not fully understood. We and many others have carried out organ ablation studies in attempts to shed light on what might be controlling the basic rhythmicity. Initially we investigated the effects of adrenalectomy, adrenalmedullectomy and hypophysectomy on rhythms in cell division, using primarily the corneal epithelium of the rodent (Fig. 11). In each case, the circadian patterns in cell division persisted, but the amplitude as was reduced compared to the controls (Scheving and Pauly, 1967). Later, in response to reports that the suprachiasmatic nuclei (SCN) disrupted or abolished rhythms of motor activity, body temperature and of several neuroendocrine functions, we evaluated the effect of SCN lesioning on the mitotic-index rhythm of the mouse cornea. We found that this procedure did not abolish the rhythm, but it did have a profound effect in reducing its amplitude and in producing a phase advance (Fig. 12). Our conclusion was that the nuclei appear to function as pace-resetter and amplifier of the rhythm (Powell et al., 1980).

We have been carrying out comprehensive, chronobiologically-controlled studies to evaluate the effect of various hormones on the DNA-synthesis rhythms in various tissues. Recently Scheving, L. A., et al. (1979, 1980) and Yeh et al. (1981) investigated the effect that epidermal growth factor

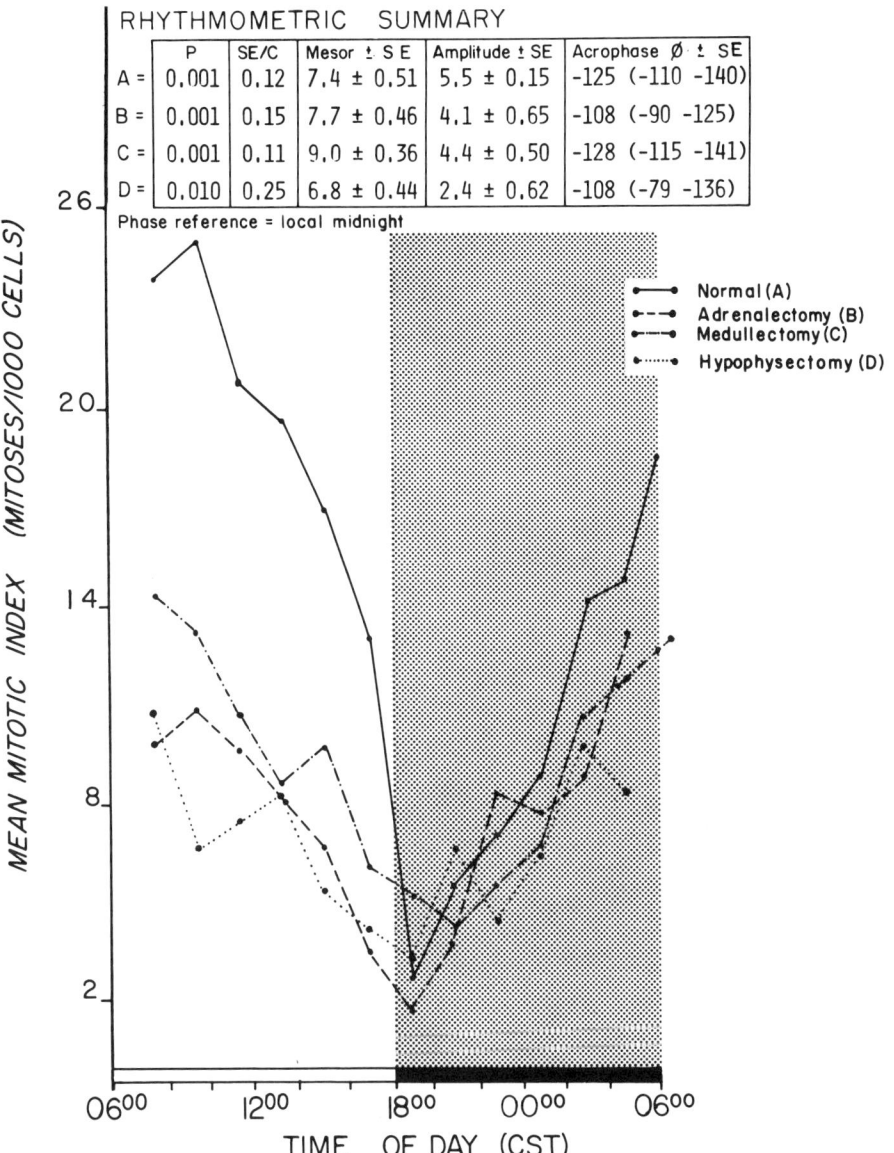

Fig. 11. Comparison of mitotic-index rhythm in corneal epithelium of normal rats with that in rats sampled three weeks after adrenalectomy, adrenal medullectomy or hypophysectomy. Adrenalectomized rats died when deprived of physiological saline; this was taken as evidence of complete adrenalectomy. For a more detailed description, see Scheving and Pauly (1967).

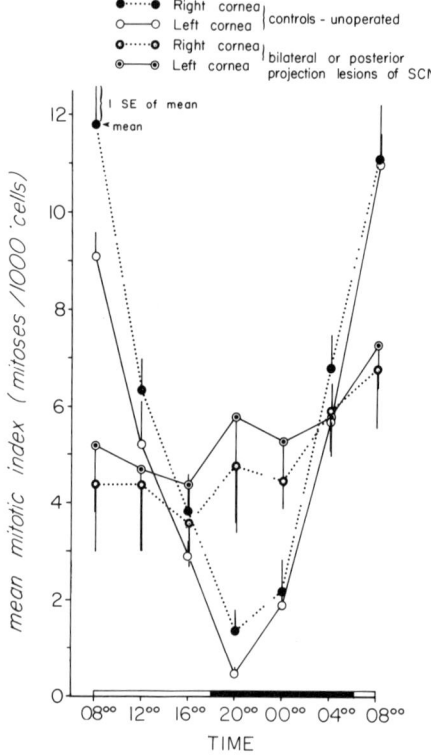

Fig. 12. Comparison of the rhythmic pattern of the mitotic index in SCN-lesioned and sham-operated mice (for description see Powell et al., 1980).

(EGF) has on (^3H)TdR incorporation into DNA in 20 different tissues. We hypothesized that EGF could be playing an important role in the control of the rhythm in cell proliferation, especially in the digestive tract (Scheving, L.A., et al., 1979). Figure 13 illustrates an example of DNA synthesis stimulation by EGF in the glandular portion of the stomach of mice 4, 8 and 12 hours after injection of EGF. Several conclusions relative to EGF have been reached. First, our results suggest that EGF plays an important role in stimulating cell-proliferation in many tissues, particularly the cornea and digestive tract. Interestingly, the

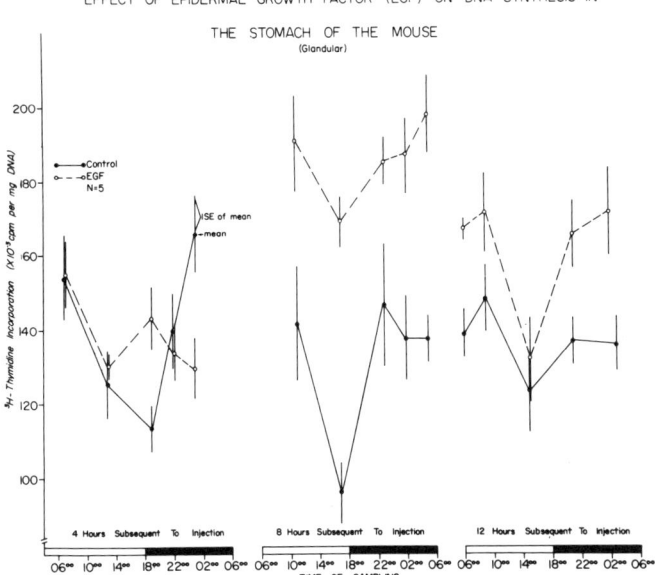

Fig. 13. Effect of epidermal growth factor (EGF) on (^{3}H)TdR incorporation into DNA of the glandular stomach is plotted versus time of sacrifice for the different injection times. Mice were killed at either 4, 8, or 12 hours after injection for each injection time (modified from data of Scheving, L. A., et al., 1980).

DNA-synthesis rhythms of two of the most responsive tissues to EGF, namely the cornea and esophagus can be dissociated in phase from one another by changing lighting schedule and feeding time, suggesting that a general systemic release of EGF is insufficient by itself to account for rhythmic variation in these tissues. Secondly, the different tissues exhibited considerable variation in their responses to EGF at 4, 8 and 12 hours after injection (EGF was injected into different groups of mice at five different time points to determine the circadian stage dependency of the effects of EGF). For example, the cornea, tongue and esophagus which were the most rhythmic tissues studied, consistently responded to EGF with increased DNA synthesis at 4, 8 and 12 hours after injection; but glandular and non-glandular stomach, colon and rectum responded only at 8 and 12 hours; the lung and aorta at 8 hours; and the liver and testes at 12 hours. Thus, the response to EGF is tissue-specific and related to time of day (circadian stage). Thirdly, there are certain circadian stages when the stimulatory effect of EGF appears to be maximally potentiated. For example, the injection of EGF at 1500 into mice standardized to an LD 12:12 cycle ($L0600D1800$) resulted in a striking generalized increase in DNA synthesis 4 hours later (at 1900), particularly in tissues such as the parotid, thymus and small intestine which at other times failed to respond normally to EGF. Interestingly, the serum concentrations of various hormones, such as insulin, gastrin and corticosterone, peak during this period of time and the rate of feeding is increased (Scheving, L. A. et al., 1980). Fourthly, our results indicate that EGF can also inhibit DNA synthesis at certain times in several tissues including the small intestine, thymus, bone marrow and spleen.

Our results also suggest that the DNA-synthesizing tissues in the digestive tract of the mouse that are the most rhythmic exhibit the greatest response to EGF. Although the relationship between rhythmicity, response to exogenous EGF, and potential to undergo transformation remains to be elucidated, the EGF cell-surface-effector system has been implicated in the in vitro chemical and viral transformation of certain cell types; this is evidenced by the fact that transformed cells, in contrast to their normal counterparts, exhibit a decreased requirement for exogenous EGF to undergo cell proliferation (Cherington et al., 1979). Also several tumor growth factors capable of interacting through the EGF-cell surface receptors have been identified in transformed cells such as SV 40 sarcoma.

EGF itself has been shown to be co-carcinogenic in the induction of certain skin tumors (Roberts et al., 1976); and recently it has been shown in rats that the removal of the submandibular gland, a major source of EGF, results in a reduction of the number of dimethylbenzanthracene-induced tumors of the colon (Li et al., 1980). This same chemical appears to be most carcinogenic at a time when the glandular content of EGF is expected, from the work of Krieger et al. (1976), to be the highest in male adult rodents. Moreover it is of interest that the induction of submandibular gland sarcomas using the same chemical has been shown to be circadian-stage dependent (Chaudhry and Halberg, 1960; Halberg, 1964) (Fig. 14).

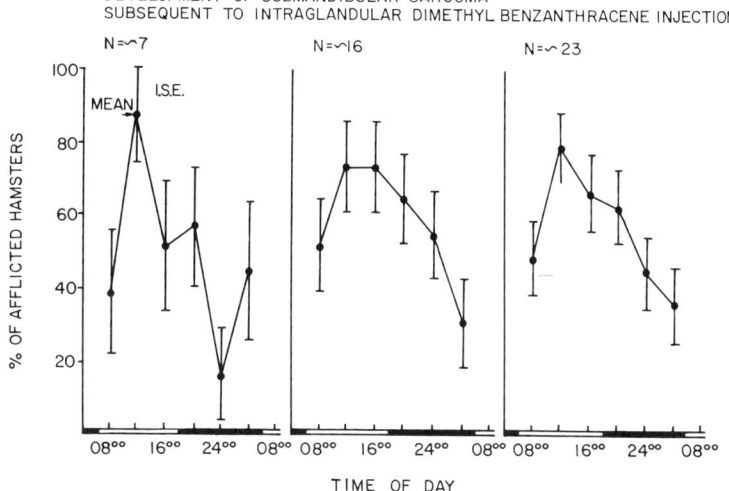

Fig. 14. Circadian variation in response to a carcinogen that was injected into the salivary glands of hamsters. Note that the highest incidence of tumor was recorded in animals injected during the light period (Chaudhry and Halberg, 1960; Halberg, 1964).

Our own studies on sialoadenectomized mice suggests that this procedure had little effect on the phasing of the rhythms but did significantly decrease over-all DNA synthesis in the liver and kidney while paradoxically increasing DNA synthesis in the colon (unpublished). Much more work needs to be done to fully resolve the role of EGF in control of cell proliferation; certainly EGF must be recognized as a powerful mitogenic agent on many tissues and is likely to be intensively investigated in the future.

We also have been interested in what effects insulin and glucagon have on cell division; our interest arose from the demonstration by Bucher et al. (1978) of the importance of EGF, insulin and glucagon in supporting hepatic cell proliferation. In summary, we have found that insulin has a stimulatory effect on the incorporation of (^{3}H)TdR into DNA in many parts of the digestive tract, especially in the stomach; but its effect is strongly circadian-stage dependent (Scheving, et al., 1981b). For example, if insulin is injected at the beginning of the dark span (animals standardized to $L^{0600}D^{1800}$) the incorporation of (^{3}H)TdR into DNA of the stomach was increased 8 hours later by 86% over controls; this increase was highly statistically significant (P<.01). At other circadian stages the response to the same dose of insulin (1 IU/25 g mouse) was less, and at some stages there was no statistically significant response. A dose-response curve indicated that as little as 0.5 units of insulin were active at certain circadian stages. The point to emphasize is that the effect of insulin on DNA synthesis is strongly circadian-stage dependent; moreover one cannot generalize about its effect from one region of the intestinal tract to another.

Although glucagon did have an effect on the incorporation of (^{3}H)TdR into DNA in some parts of the digestive tract (not the stomach), its effect was not as dramatic as sometimes seen for EGF or insulin. The effect observed for one tissue obtained from preliminary studies is shown in Figure 15. In this case it was apparent in that glucagon was capable of stimulating the incorporation of (^{3}H)TdR into DNA in the caecum of the mouse at one circadian stage but not at another. Since these preliminary data were presented at this symposium, we have found that this polypeptide is also capable of depressing DNA synthesis in the caecum if administered at still another circadian stage (data to be published elsewhere).

Clearly, as is the case *in vitro*, EGF, insulin and glucagon all have an effect on DNA synthesis, especially in the digestive tract. Further studies are needed to fully resolve the role of each and of other similar molecules. If we are ultimately to understand fully the effects that hormones, etc., have on the induction and control of cell proliferation, we cannot ignore the basic rhythmicity while seeking the mechanism.

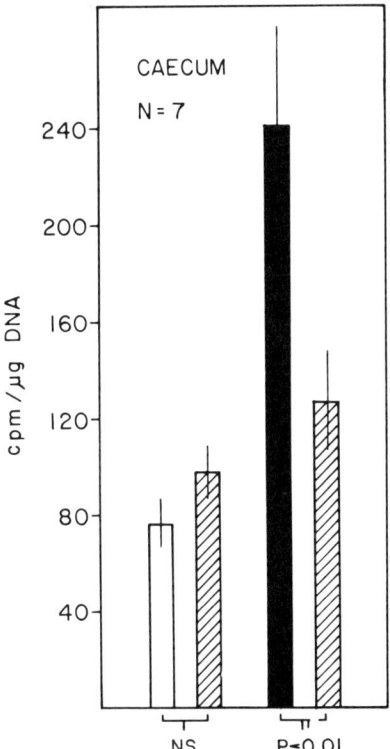

Fig. 15. The data show that glucagon significantly increases the incorporation of (^3H)TdR into the DNA of the caecum if administered at the beginning of the dark span but has no significant effect if injected toward the end of the light span. The dosage was 25 µg of glucagon in 0.2 ml of saline/ 25 g mouse (not previously published).

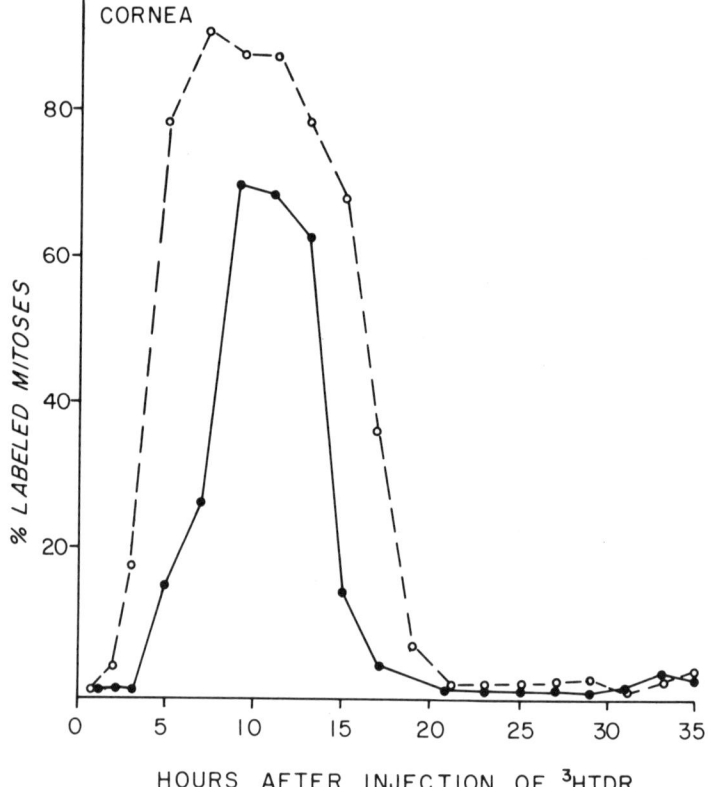

Fig. 16. The frequency of labeled mitosis method (FLM) was used to determine the duration of the S phase of the cell cycle in the epithelium of the mouse cornea. The dashed line represents the data obtained from mice that were injected with (^{3}H)TdR at 0900 and killed at frequent intervals thereafter. The solid line represents data obtained from mice that were injected with (^{3}H)TdR at 2100 and killed at frequent intervals thereafter. (For complete details see Burns and Scheving, 1975; also see text for further explanation.)

WHAT VALUE IS THERE IN RECOGNIZING RHYTHMS IN CELL DIVISION?

Examination of cell-proliferation time in many tissues has been made using the once-popular FLM method originally designed by Quastler and Sherman (1959). In fact, the intestinal epithelium, which we have demonstrated to be extremely rhythmic, was used by Quastler and Sherman because they believed that it divided randomly. Reasons for this erroneous view have been discussed (Burns and Scheving, 1975; Scheving and Burns, 1977). Obviously, this method no longer can be considered reliable for investigating cell kinetics in the in vivo system. Figure 16 illustrates why this statement is made.

We have shown that if one injects (^{3}H)TdR when the mitotic index is lowest in the corneal epithelium, about 2100, and analyzes the cell division in this tissue in a conventional manner, then G_2 + 1/2 M = 8 hours, and T_S = 5.4 hours. If (^{3}H)TdR is injected at the time of highest mitotic activity at 0900, then G_2 + 1/2 M = 4 hours and T_S = 12.2 hours (Burns and Scheving, 1975; Scheving and Burns, 1977). Tvermyr (1972) and Møller et al. (1974) have reported similar findings for the epidermis of the hairless mouse and the hamster cheek pouch, respectively. Unfortunately the FLM method is still being used and continues to generate questionable data; for examples, one need only to read the 1979 and 1980 issues of certain journals specializing in this type of data. Some workers using the FLM technique claim to sample at the same time of day; apparently they do this thinking they are avoiding the rhythm problem. Obviously from the data presented here and from the literature cited in this paper, this is no solution. To appreciate this fact, one has only to reflect for a moment on the ability of drugs and of other perturbations to phase-shift rhythms to realize that the killing of control and experimental animals at the same time of day does not necessarily produce data that reflect what is actually happening.

Any technique used in the study of in vivo cell kinetics that cannot resolve the inherent rhythmicity in cell proliferation is unreliable. We are pleased to find that Laerum and Aardal as well as Rubin, reporting at this same symposium, have demonstrated rhythms in the S, M, G_1 and G_2 phases of the cell cycle using the flow microfluorometry (FMF) method. This technique would appear to have promise in exploring further, with far less labor, cell-proliferation rhythms, at least in certain tissues (Rubin, this symposium).

Susceptibility Rhythms

Because there is a rhythm in cell proliferation, it should not be surprising that there also is a rhythm in toxicity to any drug which has as its target any specific phase of the cell cycle. Such a rhythm is illustrated in

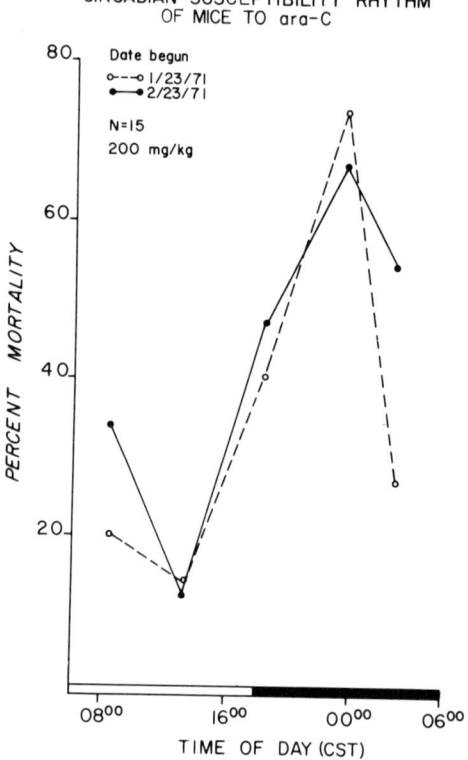

Fig. 17. Circadian-susceptibility rhythm in BD_2F1 mice to arabinosyl cytosine (ara-C) given on five consecutive days at single, defined circadian-system stages (Scheving et al., 1974b).

Figure 17; 15% of the animals treated succumbed to the toxic effects of arabinosyl cytosine (ara-C) administered at one circadian stage, whereas approximately 74% of the animals died after being given the same injection of the drug at another circadian stage (Cardoso et al., 1970; Scheving et al., 1974b). When one compares this variation in drug toxicity to the rhythm in DNA synthesis in the duodenum, it is seen that the highest mortality occurs at the time when DNA synthesis is just beginning its daily upswing (Fig. 18) (Scheving et al., 1977b). Of course we cannot be sure that this is the cause of the variation in toxicity, but it is recognized that the bone marrow and gut are two of the primary target organs of this agent.

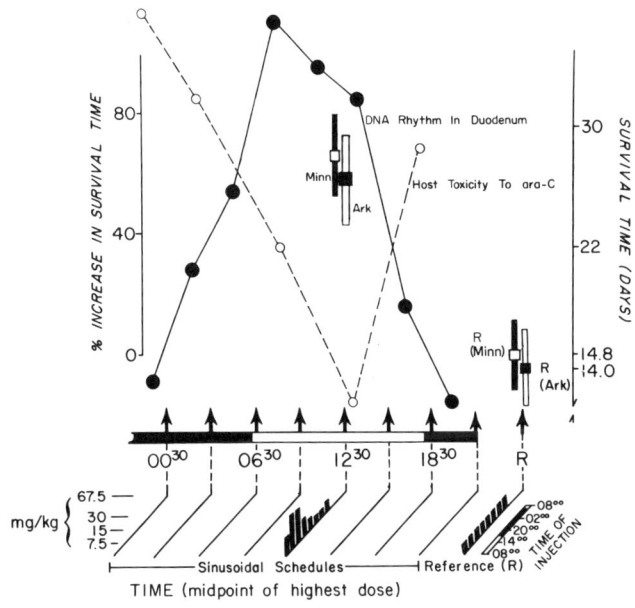

Fig. 18. Comparison of the ara-C toxicity rhythm to the rhythm in DNA synthesis in the duodenum (see text and Scheving et al., 1977b).

Because the biological system is rhythmically changing, it follows that the organism is biochemically a different entity at different circadian stages; consequently, as illustrated above for ara-C, it reacts differently to the same stimulus when applied at different times. This differential response to an identical stimulus has been confirmed repeatedly for a variety of stimuli; these include: drugs, poisons, chemical substances, physical agents such as noise and X-irradiation, and biological agents such as endotoxins (for reviews see Reinberg and Halberg, 1971; Scheving et al., 1974c; Moore Ede, 1973). Figure 19 gives additional examples of such rhythms in toxicity in animal models and Figure 20 illustrates similar susceptibility rhythms in humans (Scheving, 1980).

We and others have carried out a series of studies which suggest that such variation in susceptibility to several anti-cancer agents which are cell-cycle specific can be optimized in experimental cancer chemotherapy (Haus et al., 1972; Kühl et al., 1974; Scheving et al., 1977). For a review, see Scheving (1980) and Levi et al. (1980). Figure 21 presents one such example . In this particular case, cyclophophamide (100 mg/kg), and adriamycin (5 mg/kg) were found to be synergistic in treating mice that had been inoculated with 1x 10^5 L1210 leukemia cells four days prior to treatment. With only one course of treatment, there was a dramatic circadian variation in response as monitored by mean survival time and cure rate. The variation in cure rate (mice alive and apparently free fo disease 75 days post-tumor inoculation) as a function of treatment timing ranged from 8% t 68% in male animals standardized to 12 hours of light alternating with 12 hours of darkness. Similarly, in female mice standardized to 8 hours of light alternating with 16 hours of darkness, the cure rate ranged from 0 to 56% depending upon when the drugs were injected during the 24-hour span. No cures were obtained with either drug alone (Scheving et al., 1980). The maximum cure rate was recorded when the two drugs were administered in the early part of the dark portion of the light-dark cycle (whether 12 hours of light and 12 hours of darkness or 8 hours of light and 16 hours of darkness); maximum mortality occurred following treatment in the light span. The data also document that maximal therapeutic advantage was obtained when the two drugs were separated by 2- or 3-hour intervals and that this effect of drug sequencing is strongly circadian-stage dependent (Scheving et al., 1980).

The results in the experimental models demonstrate clearly that one should not ignore circadian variation in tolerance when dealing with chemotherapy or sequencing studies. The evidence is compelling that the temporal

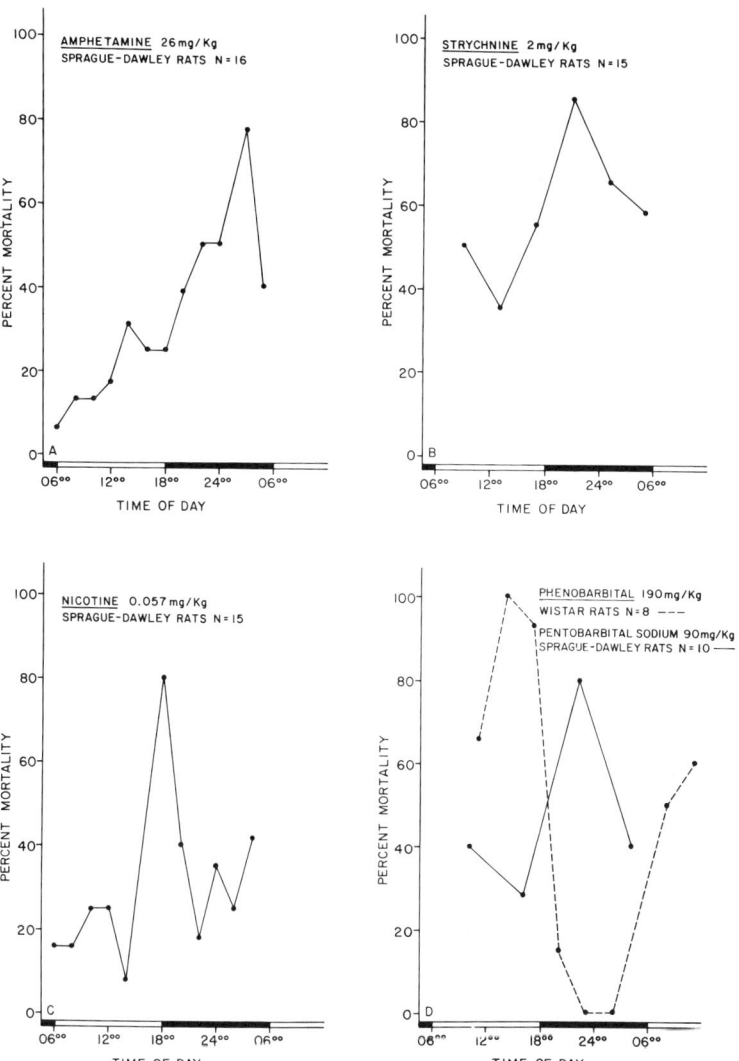

Fig. 19. Circadian-susceptibility rhythms in rodents in response to four agents, using mortality as the endpoint. (For details see Scheving and Vedral, 1966, and Scheving et al., 1968 for amphetamine data; Tsai et al., 1970 for strychnine and nicotine data; Müller, 1974, for phenobarbital sodium data.) Note in the case of the long acting barbiturate (phenobarbital) that at one circadian stage 100% of the rats survived whereas at another they all died.

organization in general carries with it significant implications, not only for cancer chemotherapy and radiotherapy, but equally important for basic research on normal as well as abnormal growth (Scheving, 1980; Haus et al., 1974).

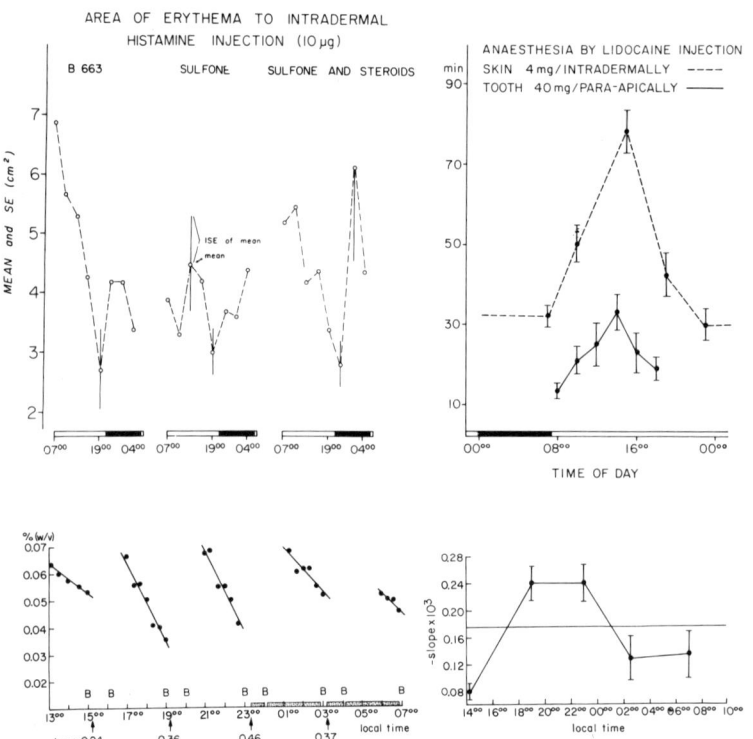

Fig. 20. Circadian-susceptibility rhythms in humans. For further details see text and Scheving et al. (1975) for histamine data; Reinberg and Reinberg (1977) for lidocaine data; and Sturtevant et al. (1978) for ethanol data. The composite graph was previously published by Scheving (1980) which also contains a more detailed explanation.

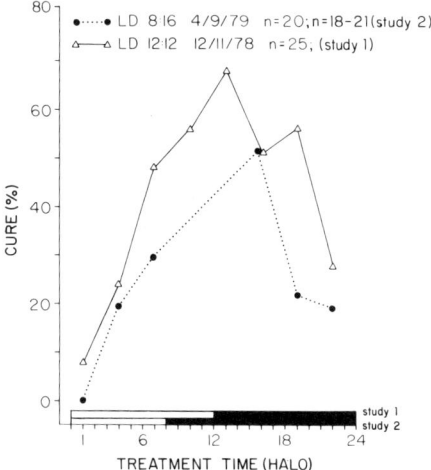

Fig. 21. Variation in cure rate of leukemic mice depending on timing of combined treatment with cyclophosphamide (CTX) and adriamycin (ADR). Study 1 (solid line) on male mice in LD 12:12; Study 2 (dotted line) on female mice in LD 8:16, HALO = hours after lights on (Scheving et al., 1980).

Perhaps we can come to better grips with the mechanism of such diseases as cancer if we consider in the first place the rhythmic nature of cell division.

We have just reached the stage where most scientists accept without question the existence of and the potential importance of such rhythms, even though some may still ignore them in experimental design. Based on the facts, we must abandon the erroneous concept that somehow sampling at the same time of day takes care of the rhythm problem (Scheving, 1980).

We hope that there will not be a similar span of 50 years before the importance of rhythmicity to cancer biology and treatment in the clinics is adequately explored. One pioneering study recently reported has shown in the clinics that kidney toxicity to cis-platinum can be dramatically reduced in cancer patients by timing the drug to body rhythms (Hrushesky et al., 1980; Levi et al., 1980). Halberg et al. (1977) in collaboration with Gupta of India (Gupta and Deka, 1975), has reported that the timing of radiotherapy in relation to a marker rhythm (tumor temperature) determines the rate of regression in peri-oral tumors. Moreover, it has been reported in an abstract that human bone-marrow toxicity to adriamycin is reduced by circadian timing (Hrushesky et al., 1981). We predict that more such clinical studies will rapidly follow.

ACKNOWLEDGEMENT

Some of the work reported here was supported by grant OH-00952 from the National Institute of Occupational Safety and Health.

REFERENCES

Andrew W, Andrew NB (1949). Lymphocytes in the normal epidermis of the rat and man. Anat Rec 104:217.
Barnum CP, Jardetzky CD, Halberg F (1957). Nucleic acid synthesis in regenerating liver. Tex Rep Biol Med 15:134.
Blumenfeld CM (1939). Periodic mitotic activity in the epidermis of albino rat. Science 90:446.
Blumenfeld CM (1943). Studies of normal and abnormal mitotic activity. II. The rate and periodicity of the mitotic activity of experimental epidermoid carcinoma in mice. Arch Path (Chicago) 35:667.
Bostroem E (1928). "Der kerbs des Menschen". Leipzig, Georg Thieme Verlag.
Broders AC, Dublin WB (1939). Rhythmiticity of mitosis in the epidermis of human beings. Proc Staff Mtg, Mayo Clinic, 14:423.
Burns ER, Scheving LE (1973). Isoproterenol-induced phase shifts in circadian rhythm mitosis in murine corneal epithelium. J Cell Biol 56:605.

Burns ER, Scheving LE (1975). Circadian influence of the waveform of the frequency of labeled mitoses and mouse corneal epithelium. Cell Tiss Kinet 8:61.

Bucher NLR, Patel U, Cohen S (1978). Hormonal factors concerned with liver regeneration. In "Hepatotrophic Factors; A Ciba Symposium", p 95.

Cameron JH (1936). The origin of new epidermal cells in the skin of normal and x-rayed frogs. J Morph 59:327.

Cardoso SS, Scheving LE, Halberg F (1970). Mortality of mice as influenced by the hour of day of drug (ARA-C) administration. Pharmacologist 12:302 (abst).

Carleton A (1934). A rhythmical periodicity in the mitotic division of animal cells. J Anat 68:251.

Chaudhry AP, Halberg F (1960). Rhythms in blood eosinophils and mitoses of hamsters pinna and pouch; phase alterations by carcinogen. J Dent Res 39:704.

Cherington PV, Smith BL, Pardee AB (1979). Loss of epidermal growth factor requirement and malignant transformation. Proc Natl Acad Sci 70:3937 (abst).

Cooper ZK, Franklin HC (1940). Mitotic rhythms in the epidermis of the mouse. Anat Rec 78:1.

Cooper ZK, Schiff A (1938). Mitotic rhythms in the human epidermis. Proc Soc Exper Biol Med 39:323.

Fisher LB (1968). The diurnal mitotic rhythm in the human epidermis. Br U Derm 80:75.

Fortuyn-van-Leyden CED (1917). Some observations of periodic nuclear division in the cat. Proc Kon Akad van Wetenschappen te Amsterdam 38:44.

Fortuyn-van-Leden CED (1926). Day and night period in nuclear division. Proc Kon Akad van Wetenschappen te Amsterdam 29:579.

Frieboes W (1920). Beitrage zur Anatomie und Biologie der Haut. Dermat Ztschr 31:57.

Gasser RF, Scheving LE, Pauly JE (1972a). Circadian rhythm in the mitotic index and in the uptake of ^{3}H-thymidine by the tongue of the rat. J Cell Physiol 80:437.

Gasser RF, Scheving LE, Pauly JE (1972b). Circadian rhythm in the cell division rate of the inner enamel epithelium and in the uptake of ^{3}H-thymidine by the root tip of rat incisors. J Dental Res 51:740.

Goodrum PJ, Sowall JG, Cardoso SS (1974). Characterization of the circadian rhythm of mitosis in the corneal epithelium of the immature rat. In Scheving LE, Halberg F, Pauly JE (eds): "Chronobiology," Tokyo: Igaku Shoin, p 29.

Gupta BD, Deka AC, Halberg E (1975). Application of chronobiology to radiotherapy of tumor of the oral cavity. Chronobiologia 2, Suppl 1:125 (abst).

Halberg F (1964). Grundlagenforschung zur Aetiologie des Karzinoms. Mkurse arztl Forbild 14:67.

Halberg F, Halberg E, Barnum CP, Bittner JJ (1959). Physiologic 24-hour periodicity in human beings and mice, the lighting regimen and daily routine. In Withrow RB (ed.): "Photoperiodism and Related Phenomena in Plants and Animals," Washington: AAAS Publ. 55, p 803.

Halberg F, Tong YL, Johnson EA (1967). Circadian system phase - an aspect of temporal morphology; procedures and illustrative examples. In Mayersbach Hv (ed.): "The Cellular Aspects of Biorhythms", Heidelberg: Springer-Verlag, p 20.

Halberg F, Johnson EA, Nelson W, Runge W, Sothern R (1972). Authorhythmometry procedures for physiologic self-measurements and their analysis. Physiology Teacher 1:1.

Halberg F, Gupta BD, Haus E, Halberg E, Deka AC, Nelson W, Sothern RB, Cornelissen G, Lee JK, Lakatua DJ, Scheving LE, Burns ER (1977). Steps toward a cancer chronopolytherapy. In: "Proceedings of XIVth International Congress of Therapeutics", Montpellier, France: L'Expansion Scientifique Francaise, p 151.

Haus E, Halberg F, Scheving LE, Cardoso S, Kühl A, Sothern R, Shiotsuka R, Hwang DS, Pauly JE (1972). Increased tolerance of leukemic mice to arabinosyl cytosine with schedule adjusted to circadian system. Science 77:80.

Haus E, Halberg F, Loken MK, Kim YS (1974). Circadian rhythmometry of mammalian radiosensitivity in Tobias CA, Todd P (eds): "Space Radiation Biology and Related Topics", New York: Academic Press, p 435.

Hrushesky W, Levi F, Halberg F, Haus E, Scheving LE, Sanchez S, Medini E, Brown H, Kennedy BJ (1980). Clinical chrono-oncology. In Scheving LE, Halberg F (eds): "Chronobiology: Principles and Applications to Shifts in Schedules", Alphen aan den Rijn, The Netherlands: NATO Advanced Study Inst Ser, Sijthoff and Noordhoff, p 513.

Hrushesky W, Halberg F, Heinlen T, Murray C, Kennedy BJ (1981). Human bone marrow toxicity of adriamycin (AD) reduced by optimal circadian timing. Cancer Res (abst). (in press).

Kanabrocki EL, Scheving LE, Halberg F, Brewer RL, Bird TJ (1974). Circadian variation in presumably healthy young soldiers. US Dept Commerce, Doc No PB228427 Nat Tech Inf Serv, P O Box 51553, Springfield, VA 22151.

Karsten G (1918). Uber Tagesteriode der kern-und zell teilung. Zeit f Bot 10:1.

Killicott WE (1904). The daily periodicity of cell division and the elongation in the root of Allium. Bull Tory Bot Club 31:529.

Killman SA, Cronkite EP, Fliedner TM, Bond VT (1962). Mitotic indices of human bone marrow cells. I. Number and cytologic distribution of mitoses. Blood 19:743.

Krieger DT, Hauser H, Liotta A, Zelenetz A (1976). Circadian periodicity of epidermal growth factor and its abolition by superior cervical ganglionectomy. Endocrinology 99:1589.

Kühl JFK, Haus E, Halberg F, Scheving LE, Pauly JE, Cardoso SS, Rosene G (1974). Experimental chronotherapy with ara-C; Comparison of murine ara-C tolerance on differently timed treatment schedules. Chronobiologia 1:316.

Laerum OD, Aardal NP (1981). Chronobiological aspects of bone marrow and blood cells. In Scheving LE, Pauly JE (eds): "Biological Rhythms in Structure and Function", New York, Alan R. Liss Inc, p 87.

Levander J (1950). On the epithelium-regeneration in the healing wounds. Acta Chir Scandianav 100:637.

Levi F, Hrushesky W, Haus E, Halberg F, Scheving LE, Kennedy BJ (1980). Experimental chrono-oncology. Principles and Applications to Shifts and Schedules", Alphen aan den Rijn, The Netherlands: NATO Advanced Study Inst Ser, Sijthoff and Noordhoff, p 481.

Li AK, Schattenkerk ME, de Vries JE, Ford WDA, Malt RA (1980). Submandibular sialoadenectomy retards dimethylhydrazine induced colonic carcinogenesis. Gastroenterology 78:1207 (abst).

Mauer AM (1965). Ciurnal variation of proliferative actifity in the human bone marrow. Blood 26:1.

Møller U, Larsen JK, Faber M (1974). The influence of injected triatiated thymidine on the mitotic circadian rhythm in the epithelium of the hamster cheek pouch. Cell Tissue Kinet 7:1007.

Moore Ede M (1973). Circadian rhythms of drug effectiveness and toxicity. Clin Pharmacol Ther 14:925.

Müller O (1974). Circadian rhythmicity and response to barbiturates. In Scheving LE, Halberg F, Pauly JE (eds): "Chronobiology", Tokyo: Igaku Shoin, p 187.

Nelson W, Zinneman H, Selden JA, Schaber K, Halberg F, Bazin H (1974). Circadian rhythm in Bence-Jones protein excretion by LOU rats bearing a transplantable immunocytoma, responsive to adriamycin treatment. Int J Chronobiology 2:359.
Ortiz-Picón JM (1934). Uber zellteilungsfrequenz und Zellteilungsrhythms in der epidermis der Maus. Ztschr f Zellforsch u Mikr Anat 19:488.
Pauly JE (1981). An introduction to chronobiology. In Scheving LE, Pauly JE (eds): "Biological Rhythms in Structure and Function", New York: Alan R. Liss Inc, p 1.
Pauly JE, Scheving LE (1964). Temporal variations in the susceptibility of white rats to pentobarbital sodium and tremorine. Int J Neuropharmacol 3:651.
Pauly JE, Burns ER, Halberg F, Tsai S, Betterton HO, Scheving LE (1975). Meal-timing dominates lighting regimen as a synchronizer of the eosinophil rhythm in mice. Acta Anat 93:60.
Phillipens KMH, Mayersbach H, Scheving LE (1977). Effects of the scheduling of meal-feeding at different phases of the circadian system in rats. J Nutr 107:176.
Powell EW, Pasley JN, Scheving LE, Halberg F (1980). Amplitude-reduction and acrophase-advance of circadian mitotic rhythm in corneal epithelium of mice with bilaterally lesioned suprachiasmatic nuclei. Anat Rec 197:277.
Quastler H, Sherman FG (1959). Cell population kinetics in the intestinal epithelium of the mouse. Exp Cell Res 17:420.
Reinberg A, Halberg F (1971). Circadian chronopharmacology. Ann Rev Pharmacol 2:455.
Reinberg A, Reinberg MA (1977). Circadian changes of duration of action of local anaesthetic agents. Naunyn-Schmiedeberg's Arch Pharmacol 297:149.
Roberts ML, Friston JA, Reade PC (1976). Supression of immune responsiveness by a submandibular salivary gland factor. Immunology 30:811.
Rubin NH (1981). Circadian stage dependence in radiation: response of dividing cells in vivo. In Scheving LE, Pauly JE (eds): "Biological Rhythms in Structure and Function", New York: Alan R Liss Inc, p 151.
Scheving LA, Yeh YC, Tsai TH, Scheving LE (1979). Circadian-phase dependent stimulatory effects of epidermal growth factor on DNA synthesis in the tongue, esophagus and stomach of the adult male mouse. Endocrinology 105:1475.

Scheving LA, Yeh YC, Tsai TH, Scheving LE (1980). Circadian-phase dependent stimulatory effects of epidermal growth factor on DNA synthesis in the duodenum, jejunum, ileum, caecum, colon, and rectum of the adult male mouse. Endocrinology 106:1498.

Scheving LE (1959). Mitotic activity in the human epidermis. Anat Rec 135:7.

Scheving LE (1976). The dimension of time in biology and medicine - chronobiology. Endeavour 35:66.

Scheving LE (1980). Chronotoxicology in general and experimental chronotherapeutics of cancer. In Scheving LE, Halberg F (eds): "Chronobiology: Principles and Applications to Shifts of Schedules", Alphen aan den Rijn, The Netherlands: NATO Advanced Study Inst Ser, Sijthoff and Noordhoff, p 455.

Scheving LE, Burns ER (1977). Some evidence that a consideration of the chronobiology of a cell cycle may improve chemotherapy. In Scharf JH, Mayersbach Hv (eds), "Die Zeit und das Leben", Nova Acta Leopoldina, 46 (225): 277.

Scheving LE, Gatz AJ (1955). Mitotic activity in the human epidermis. Anat Rec 121:363 (abst).

Scheving LE, Pauly JE (1960). Daily mitotic fluctuation in the epidermis of the rat and their relation to variations in spontaneous activity and rectal temperature. Acta Anat 43:337.

Scheving LE, Pauly JE (1967). Circadian phase relationship of thymidine-H^3 uptake, labeled nuclei, grain counts and cell division rate in rat corneal epithelium. J Cell Biol 32:677.

Scheving LE, Pauly (1971). Pitfalls awaiting those who ignore circadian time structure in experimental design: a conclusion based on experiments to test the effect of cytosine arabinoside on mitosis. Anat Rec 169:419 (abst).

Scheving LE, Pauly JE (1973). Cellular mechanisms involving biorhythms with emphasis on those rhythms associated with the S and M stages of the cell cycle. Int J Chronobiology 1:269.

Scheving LE, Pauly JE (1977). Several problems associated with the conduct of chronobiological research. In Scharf KH, Mayersbach Hv (eds), "Die Zeit und das Leben", Nova Acta Leopoldina 46 225:237.

Scheving LE, Vedral E (1966). Circadian variation of susceptibility of rat to several different pharmacological agents. Anat Rec 154:417.

Scheving LE, Vedral DF, Pauly JE (1968). Daily circadian rhythm in rats to D-amphetamine sulphate: effect of blinding and continuous illumination on the rhythm. Nature 219:621.

Scheving LE, Burns ER, Pauly JE (1972). Circadian rhythms in mitotic activity and ^{3}H-thymidine uptake in the duodenum; effect of isoproterenol on the mitotic rhythm. Am J Anat 135:311.

Scheving LE, Pauly JE, Burns ER, Halberg F, Tsai TH, Betterton HO (1974a). Lighting regimen dominates interacting meal schedules and synchronizes mitotic rhythms in mouse corneal epithelium. Anat Rec 180:447.

Scheving LE, Cardoso SS, Pauly JE, Halberg F, Haus E (1974b). Variation in susceptibility of mice to the carcinostatic agent arabinosyl cytosine. In Scheving LE, Halberg F, Pauly JE (eds), "Chronobiology", Tokyo: igaku Shoin, p 213.

Scheving LE, Mayersbach Hv, Pauly JE (1974c). An overview of chronopharmacology (a general review). J Europeen Toxicol 7:203.

Scheving LE, Enna CC, Halberg F, Jacobsen RR, Mather A, Pauly JE (1975). Mean circadian cosinors of vital signs, performance and of blood and urinary constituents in patients with leprosy. Int J Leprosy 43:364.

Scheving LE, Burns ER, Pauly JE, Tsai TH, Betterton HO, Halberg F (1976). Meal scheduling, cellular rhythms and the chronotherapy of cancer. In Koishi H (ed) "Nutrition," Kyoto: Victory-sha Press, p 141.

Scheving LE, Halberg F, Kanabrocki E (1977a). Circadian rhythmometry on 42 variables of 13 presumably healthy young men. In "12th International Conferen Proceedings," International Society for Chronobiology), Milano: Il Ponte, p 47.

Scheving LE, Burns ER, Pauly JE (1977b). Can chronobiology be ignored when considering the cancer problem? In Nieburgs HE (ed), "Prevention and Detection of Cancer. Pt 1, Prevention, Vol 1, Etiology," New York: Marcel Dekker, p 1063.

Scheving LE, Burns ER, Pauly JE, Halberg F, Haus E (1977c). Survival and cure of leukemic mice after optimization of cancer treatment with cyclophosphamide and arabinosyl cytosine. Cancer Res 37:3648.

Scheving LE, Burns ER, Pauly JE, Halberg H (1980). Circadian bioperiodic response of mice bearing a L1210 leukemia to combination therapy with adriamycin and cyclophosphamide. Cancer Res 40:1511.

Scheving LE, Pauly JE and Scheving LA (1981a). Circadian
rhythms at the cellular level. In Mayersbach Hv, Reinberg
A, Smolensky M (eds), "Chronobiology in Pharmacology and
Nutrition", New York: Springer-Verlag (in press).
Scheving LE, Pauly JE, Tsai TH and Scheving LA (1981b).
Effect of insulin on DNA labelling in gastric mucosa depends on circadian stage of mouse when drug is administered.
Anat Rec (in press).
Stalfelt MG (1921). Stud. uber die periodizitat d
Zellteilung u, sick daran anschliessend Erscheinungen.
Kunst Svenska Vetensh Hand 62:1-114.
Sturtevant RP, Sturtevant FM, Pauly JE, Scheving LE (1978).
Chronopharmacokinetics of ethanol. III. Circadian variations in rate of ethanolemia decay in human subjects.
Int J Clin Pharmacol Biopharm 16:594.
Tsai TH, Scheving LE, Pauly JE (1970). Circadian rhythm in
plasma inorganic phosphorus and sulphur of the rat, also
in susceptibility to strychnine. Jap J Physiol 20:12.
Tvermyr EM (1969). Circadian rhythms in epidermal mitotic
activity. Diurnal variations of the mitotic index, the
mitotic rate and the mitotic duration. Virchows Archiv,
Abt B, Zellpathol 2:318.
Tvermyr EMF (1972). Circadian rhythms in hairless mouse
epidermal DNA-synthesis as measured by double labelling
with H^3-thymidine (H^3TdR). Virchows Arch Abt B, Zellpath
11:43.
Yeh YC, Scheving LA, Tsai TH, Scheving LE (1981). Circadian-
phase dependent effects of epidermal growth factor on
deoxyribonucleic acid synthesis in ten different organs
of the adult male mouse. Endocrinology (in press).
Zugula-Mally ZW, Cardoso SS, Williams D, Simpson H, Reinberg
A (1979). Time point differences in skin mitotic activity
of actinic keratoses and skin cancers. Circadian reference:
plasma cortisol. In Reinberg A, Halberg F (eds), "Chronopharmacology," New York: Pergamon Press, p 399.

CIRCADIAN VARIATION IN THE PROPORTION OF CELLS IN CELL CYCLE PHASES IN HAMSTER TONGUE EPITHELIUM MEASURED BY FLOW MICRO-FLUOROMETRY*

Norma H. Rubin

Department of Human Biological Chemistry and Genetics, Division of Cell Biology, University of Texas Medical Branch, Galveston, TX 77550

INTRODUCTION

The DNA content of isolated single cells can be measured rapidly and accurately by flow microfluorometry (FMF) (Van Dilla et al., 1969; Göhde and Dittrich, 1971). In a cell population the DNA distribution can be calculated readily by computer methods enabling one to determine the proportion of cells in G_1, S, and G_2+M phases of the cell cycle.

Circadian rhythms in cell division have been reported since the beginning of the centuyy (Scheving, 1959 for review). Circadian rhythmicity has been demonstrated previously both for the uptake of tritiated thymidine and for the mitotic index in the murine tongue (Gasser et al., 1972; Burns et al., 1976; Scheving et al., 1978).

In the present study the epithelium of the hamster tongue was analyzed by FMF to determine if a circadian rhythm was present in the proportions of cells in the phases of the cell cycle.

MATERIALS AND METHODS

Fifty-four male golden Syrian hamsters (Mesocricetus auratus), about 15 weeks of age, were used in this study.

---

* This work was supported by ACS Grant 2-19511-477004-10 to N.H.R. and UTMB Cancer Center Core Grant DHHS 5P30 CA 17701-06.

They were housed 10 to a cage; and for four weeks prior to the experiment, they were standardized to an artificial light-dark cycle with light from 0600 to 2000 (LD 14:10). Food and water were available ad libitum.

Animals were killed by cervical dislocation every three hours for 27 hours in order to note the effect of "time of day" on the proportion of cells in the phases of the cell cycle. At 0900 on 13 November 1980, the first group of six animals was killed. The tongues were removed and placed in 0.5% acetic acid. Three hours later at 1200, and at three-hourly intervals thereafter until 0900 on 14 November 1980, a similar group of animals was killed. After treatment for 24-48 hours at 4°C in 0.5% acetic acid, the epithelium of the tongue could be easily removed. The isolated epithelium from each animal was minced finely and then stirred for 10 minutes at room temperature with 0.5% pepsin, which dissociates the cells and results in individual nuclei without cytoplasm. After filtration through several layers of cheesecloth to remove undissociated segments, the filtrate (in conical centrifuge tubes) was fixed with 100% ethanol to a final concentration of 70% ethanol and then refrigerated at least 48 hours to allow sedimentation of the cellular elements.

To prepare a sample for FMF analysis, a 0.5-1 ml aliquot of the cells was removed, treated with several drops of 0.5% pepsin, several drops of 0.1% RNase, and 1-2 ml of ethidium bromide-mithramycin stain (Crissman and Tobey, 1974). After 25 minutes at room temperature, the total fluorescence per nucleus in the sample was measured in a flow microfluorometer (Coulter TPS-1, Coulter Electronics, Inc.) and the combined results plotted as a DNA histogram. The number of nuclei fluorescing with a G_1 amount of DNA constituted the larger peak, while a smaller peak indicated the nuclei with twice that amount of DNA, or G_2 cells. Mitotic cells are probably lost by this technique, since the pepsin pretreatment destroys the cell membrane. Nuclei with increasing amounts of fluorescence between G_1 and G_2 represented the fraction of cells in S phase. By the method of Gusman (1978) the percentage of cells in the G_1, S, and G_2+M phases of the cell cycle was calculated. A total of 15,000-20,000 cells was analyzed per sample; for each of the nine sample times, a mean and standard error were calculated for each phase of the cell cycle. Analysis of variance was performed as indicated.

RESULTS

The percentage of cells in G_1, S, and G_2+M all showed a statistically significant circadian variation (Table 1). Mean percentages for G_1 phase ranged from 81.9% and 82.0% at 0900 on 13 November and 14 November respectively, to 93.4% at 1800 (analysis of variance, $p<0.001$). Mean percentages for S phase ranged from 12.7% and 12.6% at 0900 to a low of 4.4% at 1800 ($p<0.001$). Mean percentages for G_2+M phase ranged from 5.4% and 5.5% at 0900 to 2.2% at 1800 ($p<0.001$).

Table 1. Mean ± Standard Error of Proportion of Cells in Phases of the Cell Cycle

TIME	G_1	S	G_2+M
0900	81.9±0.2	12.7±0.2	5.4±0.1
1200	85.7±0.6	9.8±0.4	4.5±0.3
1500	87.2±1.7	8.1±0.9	4.7±1.2
1800	93.4±0.2	4.4±0.3	2.2±0.2
2100	86.4±0.8	9.9±0.6	3.7±0.3
2400	87.6±0.7	9.3±0.6	3.0±0.3
0300	84.9±0.5	11.5±0.4	3.6±0.3
0600	83.3±1.0	12.0±0.6	4.7±0.6
0900	82.0±0.8	12.6±0.3	5.5±0.5

Figure 1 illustrates the variation around a 24-hour mean of the fraction of cells in the phases of the cell cycle. The G_1 phase varied from 95.0% (0900) to 108.0% (1800) of the 24-hour mean. The S phase varied from 45.6% (1800) to 130.0% (0900) of the 24-hour mean, whereas the G_2+M phase varied from 54.2% (1800) to 136.0% (0900).

DISCUSSION

Numerous workers have documented that cell division in the alimentary tract is extremely rhythmic (Sigdestad et al., 1969; (Scheving et al., 1978; Hamilton, 1979). The results of this experiment, although preliminary, further document previously reported circadian rhythms in cell division. The data, when plotted as a function of mitoses or DNA synthesis versus time, show a prominent approximately sinusoidal curve

(Scheving and Pauly, 1973). The percentage of cells in S phase varied almost three-fold, while the percentage of cells in G2+M varied 2-3 fold.

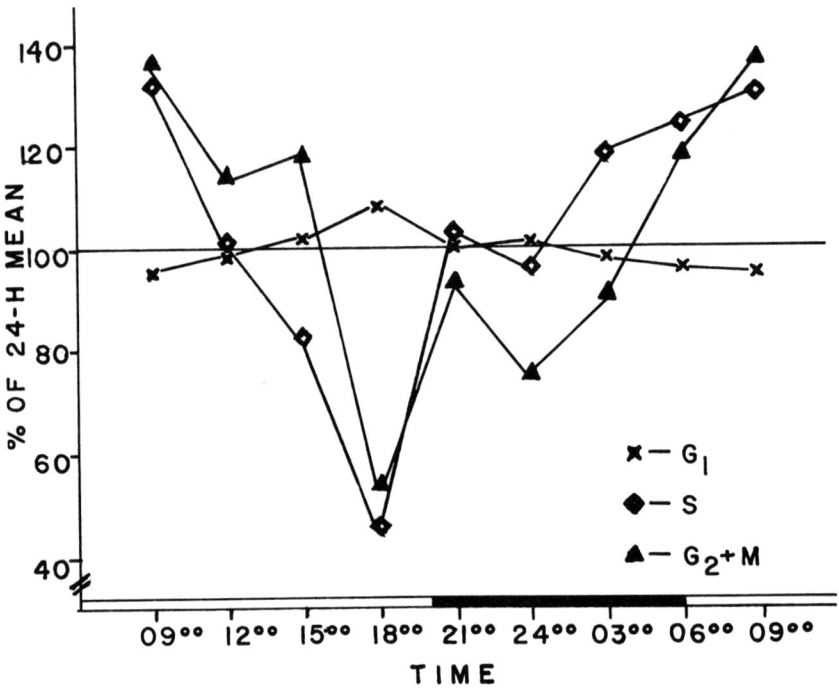

Fig. 1. Variation around the 24-hour Mean of the Fraction of Cells in the Phases of the Cell Cycle. Light-dark indicated on abscissa.

There was a perplexing coincidence of phasing of the rhythms in S and G_2. This finding would not have been anticipated from in vitro cell-kinetics studies. In such experiments it is possible to calculate specific cell-cycle transit times, and DNA synthesis regularly precedes mitosis by a measurable span of time. However, the similarity of phasing of S and M has been reported previously in vivo in chronobiologically oriented studies by Chiakulas and Scheving (1966) on larval urodele epidermis, by Scheving and Pauly (1967) on the murine corneal epithelium, by Sigdestad et al.

(1969) on the murine gut, and by Tvermyr (1972) on skin from hairless mice. Moreover, Burns and Scheving (1972) reported that with the labeled-mitosis method, cell-cycle parameters showed a significant circadian variation. They noted variations in length of S phase of 12 hours or 5.5 hours, and G_2 phase of 8 hours or 4 hours. Perhaps this variability of transit times has a bearing on the apparent coincidence in phasing of S and G_2 in this experiment.

The relatively new technique of flow microfluorometry offers several advantages over conventional methods of analyzing cell-kinetic parameters. Among these are the increased quantitative and statistical accuracy and the elimination of exposure of the animal to radioactive isotopes or mitotic-spindle poisons. This technique is especially useful for studying circadian rhythms in cell division (Thorud et al., 1978; Møller and Larsen, 1978, 1979; Clausen et al., 1979) and will find application in monitoring the effects of chronochemotherapy on normal proliferative tissues and on tumor activity.

REFERENCES

Burns ER, Scheving LE (1975). Circadian influence on the wave form of the frequency of labeled mitosis in mouse corneal epithelium. Cell Tissue Kinet 8:61.

Burns ER, Scheving LE, Fawcett DF, Gibbs WM, Galatzan RE, (1976). Circadian influence on the frequency of labeled mitoses method in the stratified squamous epithelium of the mouse esophagus and tongue. Anat Rec 184:265.

Chiakulas JJ, Scheving LE (1966). Circadian phase relationships of ^{3}H-thymidine uptake, numbers of labeled nuclei, grain counts and cell division in larval urodele epidermis. Exp Cell Res 44:256.

Clausen OPF, Thorud E, Bjerknes R, Elgjo K (1979). Circadian rhythms in mouse epidermal basal cell proliferation. Cell Tissue Kinet 12:319.

Crissman HA, Tobey RA (1974). Cell cycle analysis in 20 minutes. Science 184:1297.

Gasser RF, Scheving LE, Pauly JE (1972). Circadian rhythms in the mitotic index of the basal epithelium and in the uptake rate of H^3-thymidine by the tongue of the rat. J Cell Physiol 80:439.

Göhde W, Dittrich W (1971). Impulsfluorometrie, ein neuartiges Durchflussverfahren zur ultraschnellen Mengenbestimmung von Zellinhaltsstoffen. Acta histochem Suppl 10:429.

Gusman L (1978). In Lutz D (ed): Pulse-cytophotometry. Mathematical modeling and analysis of flow microflurometry DNA distributions. Ghent: European Press.

Hamilton E (1979). Diurnal variation in proliferative compartments and their relation to cryptogenic cells in the mouse colon. Cell Tissue Kinet 12:91.

Møller U, Larsen JK (1978). The circadian variations in the epidermal growth of the hamster cheek pouch. Cell Tissue Kinet 11:405.

Møller U, Larsen JK (1979). DNA flow cytometry of isolated keratinized epithelia: A methodological study based on ultrasonic tissue disaggregation. Cell Tissue Kinet 12:203.

Scheving LE (1959). Mitotic activity in the human epidermis. Anat Rec 135:7.

Scheving LE, Pauly JE (1967). Circadian phase relationships of thymidine-^{3}H uptake, labeled nuclei, grain counts and cell division rate in rat corneal epithelium. J Cell Biol 32:677.

Scheving LE, Pauly JE (1973). Cellular mechanisms involving biorhythms with emphasis on those rhythms associated with the S and M stages of the cell cycle. Int J Chronobiol 1:269.

Scheving LE, Burns ER, Pauly JE, Tsai T-H (1978). Circadian variation in cell division of the mouse alimentary tract, bone marrow and corneal epithelium. Anat Rec 191:479.

Sigdestad CP, Bauman J, Lesher S (1969). Diurnal fluctuation in the number of cells in mitosis and DNA synthesis in the jejunum of the mouse. Exp Cell Res 58:159.

Thorud E, Clausen OPF (1978). In Lutz D (ed): Pulse-cytophotometry. The effects of bleomycin on murine epidermal cell kinetics. Ghent: European Press, p 553.

Van Dilla MA, Trujillo TT, Mullaney PF, Coulter JR (1969). Cell microfluorometry: A method for rapid cell measurement. Science 163:1213.

CHRONOBIOLOGICAL ASPECTS OF BONE MARROW AND BLOOD CELLS

Ole Didrik Lærum, MD and Nils Petter Aardal, MD

The Gade Institute, Department of Pathology
University of Bergen
5016 Haukeland Hospital, NORWAY

From a morphological point of view the bone marrow shows a high degree of constancy, including the architecture as well as the relative composition of the different cell types. In spite of this, haematopoietic tissues have a very rapid cell turn-over. Multipotent stem cells give rise to committed stem cells of the different cell compartments, followed by maturation and cell division until the main end products, granulocytes, monocytes, erythrocytes, lymphocytes and platelets leave the marrow cavity (for review, see e.g. Robinson and Mangalik, 1975). The rapid cell proliferation in the bone marrow is, however, not constant, but is known to undergo rhythmic variations. This includes circadian as well as circannual types of rhythms. In the present article we shall show some examples of such rhythms.

GENERAL

Stem cells: Stoney et al. (1975) found that the number of multipotent stem cells (CFU_S) in the marrow of mice underwent diurnal variations with maximum (acrophase) at 06.00.

Erythropoiesis: Dörmer et al. (1970) showed that DNA synthesis in erythropoietic cells of mice as measured by ^3H-thymidine labelling index underwent diurnal variations. Maximal values were found at 03.00. Sharkis et al. (1969) found a similar variation in mitotic activity of erythropoietic cells, although males had different maxima from females, i.e. at 12-16.00 versus 20-24.00. On the other hand, Mauer (1965) was unable to find any diurnal variations of ^{3}H-Tdr labelling index in erythropoiesis of humans, although labelling index of myelopoietic cells as well as mitotic index in whole marrow underwent such variations.

McKee et al. (1974) found diurnal variations of reticulocyte counts in peripheral blood with maximum at 20.00. Similarly, Nechaev et al. (1977) found circannual variations in erythrocyte aminohydrolase over two years. Erythropoietin shows daily as well as circannual variations, where maxima have been found in January and September (Mc Kee 1974).

Myelopoiesis: In humans Mauer (1965) found that the ^{3}H-Tdr labelling index of myelopoiesis was highest in the morning and first part of the day, and lowest at midnight. Similarly, the mitotic index tended to be highest at midnight and lower in the morning. Sharkis et al. (1971) found diurnal variations of mitotic index in myelopoietic cells with maximum at 16.00 and minimum at 20.00. Similar variations were also found in lymphocytes and erythropoietic cells, although the fluctuations were only significant in males.

On the other hand Pizzarello and Witcofsky (1970) found that the total ^{3}H-Tdr labelling index in mice was maximal at 22.00 and mitotic index at 02.00. Burns et al. (1979) found maximum uptake of ^{3}H-thymidine at 05.00, while Scheving et al. (1978) found maximum in the middle of the day. In rat bone marrow, Halberg et al. (1973) found maximum ^{3}H-Tdr uptake at 18.00 and minimum at midnight. The difference between minimum and maximum was more than two times.

In peripheral blood, Halberg et al. (1973) reported a maximal number of granulocytes in the middle of the day, while Hume et al. (1975) found the acrophase at early evening with an amplitude of ± 2 times. In rabbits such variations were only found in males (Fox and Laird, 1970) where the maximal number was observed at 20.00. Lymphocytes had the maximum at 04.00. Halberg (1954) found diurnal variations of eosinophil granulocytes with high values in the day and low at night.

In other words, all the different compartments of the bone marrow show circadian variations, although the time of peak values differs from investigation to investigation and also seems to be dependent on the type of animal.

EXPERIMENTAL INVESTIGATIONS

Since diurnal variations of cell proliferation in the bone marrow had mainly been observed in male animals and only occasionally in females (see Sharkis et al. 1971 and 1974), it was of interest to investigate the female bone marrow in more detail in order to see if there were reproducible rhythms. It was also of interest to determine at what part of the day the maximal values occurred. Since rhythmic variations have been found in multipotent stem cells (Stoney et al. 1975), we have investigated if this also applies to the stem cells committed for myelopoiesis, i.e. CFU_C.

Bone marrow cell proliferation: For this purpose we used a new method which gives automatic measurement of cell cycle distribution at a very high rate, i.e. by means of flow cytometry. Female C_3H mice, weight 20-24g, were kept in cages of 5 each with 12 hours light (07.00-19.00) and 12 hours darkness. Another group of animals was investigated after 6 weeks of exposure to continuous light. The animals were killed by cervical dislocation at 3 hour intervals, and one femur from each animal was flushed with 5 ml of phosphate buffered saline (PBS). The cells were spun down gently and resuspended in a mixture of ethidiumbromide, detergent and RNAse as described by Vindeløv (1978). This method disrupts the cells and stains specifically double-stranded DNA in the isolated nuclei. Relative amount of DNA per cell was measured in an Ortho model H50 flow cytometer, which gives a plot of the relative DNA content per cell, altogether 50-100.000 per sample. The relative fraction of G_1, S-phase cells and cells in G_2 was determined by a planimetric method. Total number of bone marrow cells per femur was measured with a Coulter counter, and the total number of cells in the different cell cycle phases was estimated per femur.

The total number of nucleated cells per femur showed a biphasic circadian variation with maxima at 10.00 and 01.00 at 12 hours light/darkness regimen (in July). Similar variations were found at continuous artificial light exposure (in May).

The relative numbers of cells in DNA synthesis phase and G_2 + mitotic phase showed a tendency to diurnal variations, although the variations were not very strong. However, when calculated as absolute numbers per femur, significant diurnal variations were found. The strongest fluctuations were found in May when the mice had been exposed to continuous light, although similar variations were present also in July under ordinary lighting regimen. In both cases the maximal proliferative activity was found in the morning and first part of the day (Fig. 2).

Before the onset of maximal DNA synthesis an increase in number of cells in early S-phase could be seen, and thereafter the rest of S-phase gradually increased (Fig. 3).

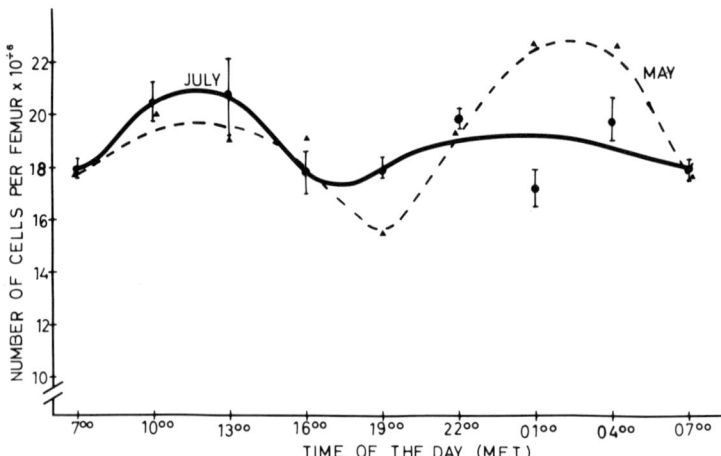

Fig. 1. Diurnal variations in total number of nucleated cells per femur in female mice. —— Mice on 12 hours light/darkness regimen (07.00 - 19.00) measured in July; each point is the mean of 3-6 mice ± S.E. --- Mice exposed to continuous light for 6 weeks, measured in May; each point is the mean of 2 mice.

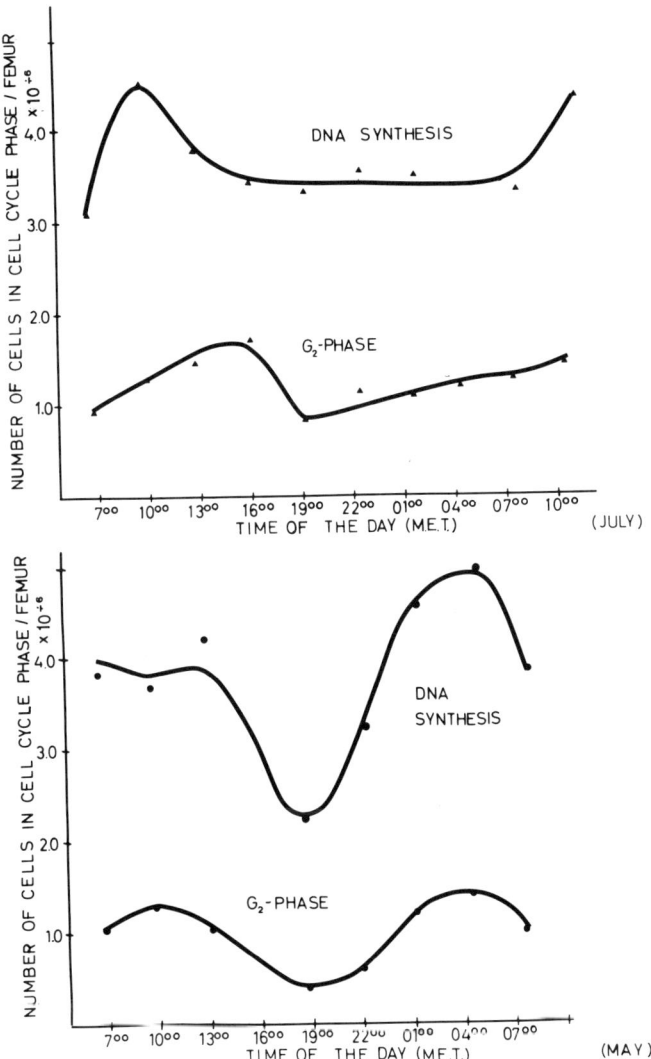

Fig. 2. Diurnal variations of total numbers of cells in DNA synthesis and G_2 phase per femur. Upper part: mice on 12/12 hours light/darkness regimen. Lower part: mice exposed to continuous light. Each point is the mean of 2-6 mice. Note the phase difference.

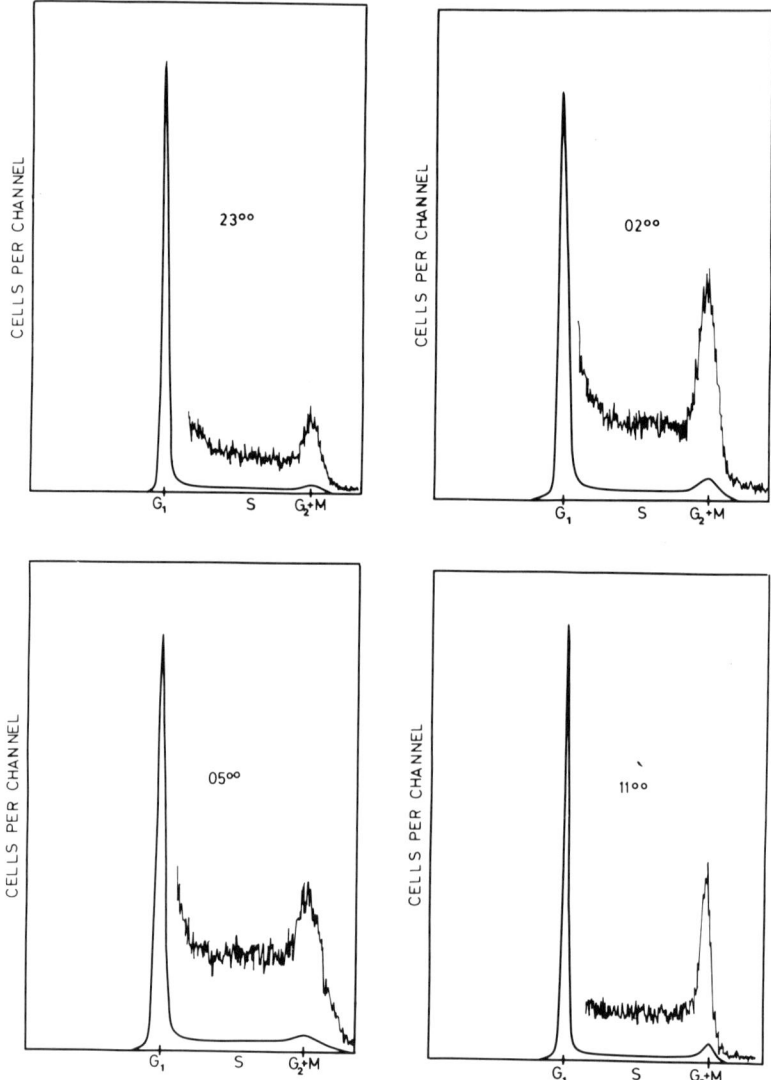

Fig. 3. Relative DNA distribution curves showing fractions of cells in different cell cycle phases at different times of the day. The curves are from the mice exposed to continuous light as shown in Fig. 2. The region of S and G_2+M phase is magnified by 10 times and shows how the period of increased cell proliferation is preceded by an increase of cells in early S-phase (at 23.00 and 02.00).

Agar colony formation: The number of colony forming units (CFU_C) per 100.000 nucleated cells was measured between 10.00 and 12.00 for 4 succeeding years. The culture medium was a modified McCoy's 5A medium (Robinson 1972) with 25% fetal calf serum. As source of colony stimulating activity (CSA), 1. pooled serum from endotoxin injected white mice of our own breeding following the method described by Quesenberry et al. (1972) and 2. lung conditioned medium from endotoxin injected mice as described by Sheridan and Metcalf (1973) was used. A one-layer method was used for the culture of bone marrow cells in agar (Bradley and Metcalf 1966). 1×10^5 cells were added to 2.0 ml 0.3% agar (Difco Bacto agar) in 35 mm culture dishes (Nunc Plastic). The dishes were incubated for 37 °C for 7 days in a humidified atmosphere of 6% CO_2 in air. The number of colonies per plate was counted after seven days (> 50 cells). As shown in Fig. 4 circannual variations were found with maxima in spring and autumn and minima in summer and winter.

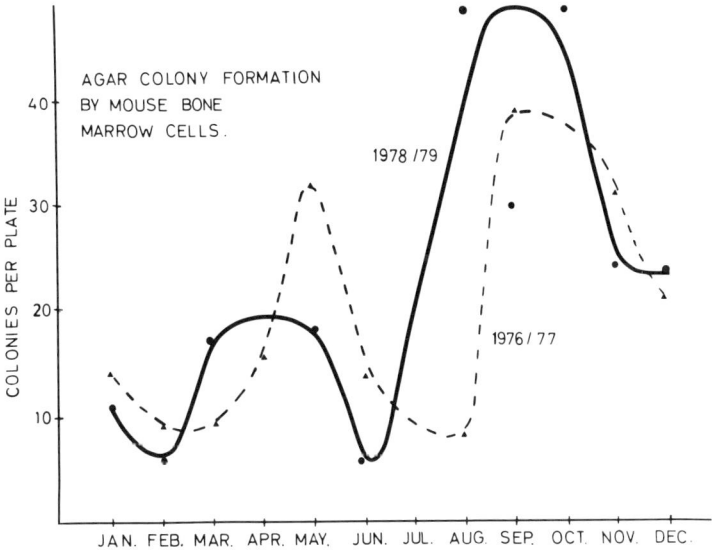

Fig. 4. Seasonal variations in numbers of colony forming cells (CFU_C) in bone marrow of female C_3H-mice. Each point is the mean of 5-50 plates. Data from two and two years are added together since they showed the same pattern. The differences between minima and maxima are significant ($p < 0.01 - 0.05$).

DISCUSSION

The determinations of the cell cycle distribution by flow cytometry gives a rapid and very precise measure of proliferative activity, which at high resolution as in the present case (1-2% coefficient of variation) also gives an indication of flux of cells through different parts of the DNA synthetic phase (Fig. 3). It shows that the bone marrow proliferation undergoes strong and reproducible circadian variations also in female mice. Even at continuous exposure to light such variations are present, although there was a phase change of peak value.

However, the present investigation does not show the relative contribution of the different parts of the bone marrow cell population, since this would require differential staining of the cells. On the other hand, the method is so rapid and reliable that it makes sequence sampling of cells rather easy over a longer time period.

In addition it is shown that the stem cells committed for myelopoiesis undergo circannual variations. Strong individual variations of the number of erythropoietic precursor cells in peripheral blood (Meytes et al. (1980) as well as of myelopoietic stem cells (Ponazzi et al. 1979) have earlier been observed, although no specific rhythmic pattern was observed in those cases. Since the mice in our case were exposed to natural light, it is tempting to suggest that seasonal light changes might have an importance for the variations observed.

In conclusion the bone marrow is a rather labile organ with a high proliferative activity which undergoes rhythmic biological variations.

Acknowledgements: This research was supported by the Norwegian Cancer Society. We thank Mrs. Gro Olderøy and Mrs. Wenche Steinsund for expert technical assistance.

REFERENCES

Aardal, NP, Laerum OD, Paukovits WR, Maurer HR (1977). Inhibition of agar colony formation by partially purified granulocyte extracts (chalone). Virchows Arch B Cell Path 24:27.

Bradley TR, Metcalf D (1966). The growth of mouse bone marrow cells in vitro. Aust J exp Biol med Sci 287:300.

Clark RH, Korst DR (1969). Circadian periodicity of bone marrow mitotic activity and reticulocyte counts in rats and mice. Science 166:236.

Dörner P, Schmolke W, Muschalik P, Brinkman W (1970). Die DNS-Synthesegeschwindigkeit im Verlaufe der DNS-Synthesephase von Erythroblasten der Maus in vivo. Beitr Path Bd 141:174.

Fox RR, Laird CW (1970). Diurnal variations in rabbits: Hematological parameters. Amer J Physiol 218:1609.

Halberg, F, Visscher MB, Bittner JJ (1954). Relation of visual factors to eosinophil rhythm in mice. Amer J Physiol 179:229.

Halberg F, Haus E, Cardoso SS, Scheving LE, Kühl JFW, Shiotsuka R, Rosene G, Pauly JE, Runge W, Spalding JF, Lee JK, Good RA (1973). Towards a chronotherapy of neoplasia: Tolerance of treatment depends upon host rhythms. Experienta 29:909.

Mauer AM (1965). Diurnal variations of proliferative activity in the human bone marrow. Blood 26:1.

McKee LC, Ensign Johnson L, Lange RD (1974). Circadian variation in reticulocyte counts and immuno-detectable erythropoietin titers (37997). Proc Soc exptl Biol Med 145:1284.

Meytes D, Ma A, Powell WB, Ortega JA, Shore NA, Dukes PP (1980). Constancy of erythroid burst forming unit (BFU_E) levels in the blood of hematologically normal individuals. Exp Hemat 8:641.

Nechaev A, Halberg F, Mittelman A, Tritsch GL (1977). Circannual variation in human erythrocyte adenosine aminohydrolase. Chronobiologia 4:191.

Pizzarello DJ, Witcofski RL (1970). A possible link between diurnal variations in radiation sensitivity and cell division in bone marrow of male mice. Radiology 97:165.

Ponassi A, Morra L, Bonanni F, Molinari A, Gigli G, Vercelli M, Sacchetti C (1979). Normal range of blood colony-forming cells (CFU-C) in humans: Influence of experimental conditions, age, sex, and diurnal variations. Blut 39:257.

Quesenberry PJ, Morley A, Stohlman F, Jr, Rickard K, Howard D, Smith M (1972). Endotoxin and colony stimulating factor. In van Bekkum DW, Dicke KA (eds): "In vitro culture of haemopoietic cells," Rijswijk: The Radiobiological Institute, TNO, p 48.

Robinson WA (1972). Demonstration of human bone marrow cultures using peripheral blood cells as feeder layer. In van Bekkum DW, Dicke KA (eds): "In vitro culture of haemopoietic cells," Rijswijk: The Radiobiological Institute, TNO, p 456.

Robinson WA, Mangalik A (1975). The kinetics and regulation of granulopoiesis. Seminars in Hematology 12:7.

Scheving LE, Burns ER, Pauly JE, Tsai T-H (1978). Circadian variation in cell division of the mouse alimentary tract, bone marrow, and corneal epithelium, and its possible implication in cell kinetics and cancer chemotherapy. Anat Res 191:479.

Sharkis SJ, LoBue J, Alexander P, Jr, Rakowitz F, Weitz-Hamburger A, Gordon AS (1971). Circadian variations in mouse hematopoiesis. II. Sex differences in mitotic indices of femoral diaphyseal marrow cells (35936). Proc Soc exptl Biol Med 138:494.

Sharkis SJ, Palmer JD, Goodenough J, LoBue J, Gordon AS (1974). Daily variations of marrow and splenic erythropoiesis, pinna epidermal cell mitosis and physical activity in C57Bl+6J mice. Cell Tissue Kinet 7:381.

Sheridan JW, Metcalf D (1973). A low molecular weight factor in lung-conditioned medium stimulating granulocyte and monocyte colony formation in vitro. J Cell Physiol 81:11.

Stoney PJ, Halberg F, Simpson HW (1975). Circadian variation in colony-forming ability of presumably intact murine bone marrow cells. Chronobiologia 2:319.

Vindeløv LL (1977). Flow microfluorometric analysis of nuclear DNA in cells from solid tumours and cell suspensions. Virchows Arch B Cell Path 24:227.

CIRCADIAN RHYTHMIC VARIATIONS OF THE RELATIVE NUMBER OF
BINUCLEATED LIVER CELLS IN RATS

K. M. H. Philippens, S. Röver, J. Abicht

Department of Anatomy, Med. Hochschule,
Hannover, W.-Germany

INTRODUCTION

The circadian (~24h) rhythmic organization of liver function has been well documented in the past three decades (13). As an integrated part of it, several nucleic activities of rat and mouse hepatocytes reveal distinct periodic variations (fig. 5). Unlike such rhythms as the waves of high and low mitotic activity and DNA synthesis, the occurrence of within-day changes of binucleate-cell (BLC) formation in liver parenchyme (5,6,29) has been investigated less extensively. As with many other functions in the animal organism, the rhythmic variation of various liver activities are influenced, directly or indirectly, by the environmental light-dark cycle, which is considered the primary or dominant time cue or "Zeitgeber". For example, the systematic 24-hour fluctuation of hepatic protein and glycogen concentration as well as of some enzyme activities has been effectively synchronized to experimentally altered light cycles (21). Since changing BLC rates closely correlate with changing conditions of metabolic load (4,17,20), the study reported here was undertaken to investigate the synchronizing effect of phase-shifted, light-dark cycles on liver cell binuclearity.

MATERIAL AND METHODS

Animals: Sixty-seven male Wistar rats, TNO substrain (TNO = Organisatie voor Toegepast Natuurwetenschappelijk Onderzoek, Zeist, The Netherlands) were maintained for 7 weeks under constant environmental conditions (lighting regimen, temperature, rel. humidity) with food (S-RMH standard laboratory chow, Hope Farms, Woerden, The Netherlands) and water available ad lib.

Lighting schedule: One group of 36 animals (subdivided into 9 subgroups of 4 animals each) served as the LD (12:12)-control group, receiving light from 06.00 - 18.00. The remaining 31 animals (subdivided into 8 subgroups of 3 - 4 animals each) were subjected to phase-shifted, light-dark (12:12) cycles (= ΔϕLD (12:12)) (fig. 1) by the use of environmental standardization boxes.

Sampling: On the day of investigation the animals each weighed 250 ± 15 grams. In the control group sampling was carried out at 3-hour intervals during an uninterrupted 24-hour period; in the group adapted to altered lighting conditions, it was done at 30-minute intervals during a 3.5-hour time span (fig. 1). In both groups sampling was started at 10.00 on the same day. The animals were killed by decapitation and subsequent bleeding. 7 μm-thin, Carnoy-fixed and Feulgen-stained sections were prepared for light-microscopic investigation of the livers.

Methods: Gross counts of the binucleate parenchymal cells (n) of each liver were determined in the central portions of 2 different sections (magnif. = 400x). The determi-

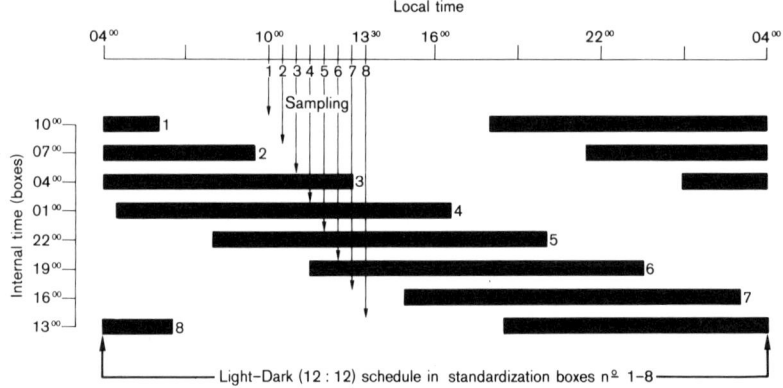

Fig. 1: Staggered phase shifts of (12:12) light-dark schedule. Black bars indicate dark spans. Sampling at 30-min. intervals from fully adapted animals simulates conventional sampling from LD (12:12)-standardized animals over full 24-h period.

nations were randomly distributed throughout the liver lobules. In each liver a total number of 7400 parenchymal cells was investigated. Estimation of the real number ($\bar{m}$) was done by Pfuhl's (1930) correction of the gross count. For this purpose the centerpoint distance of twin nuclei in binucleated cells (200 per liver) was measured by microprojection (magnif. = 4200x) of the same 2 tissue sections. The measurements were performed with an optoelectronic device attached to a microprocessor. It is obvious that these corrections cannot compensate for those binucleate cells counted as mononucleated, because by tissue sectioning they have been deprived of one of their two nuclei.

RESULTS

$\bar{n}$ (gross count of binucleated hepatocytes): In the animals living under standard LD(12:12), the 24-hour mean value of BLC counts was 14.63%. At the different time points $\bar{n}$ ranged from 8.51% (± 0.96, S.E.) during the light phase (at 10.00 at the beginning of the experiment) to 18.02% (± 1.10, S.E.) during the dark phase, showing a regular and distinct rhythmic pattern (fig. 2, table 1).

In the group maintained under phase-shifted LD (12:12) cycles, the variations of $\bar{n}$ followed basically the same pattern with the same phasing as the one described above (fig. 2, table 2). However, in this group at each time point $\bar{n}$ was some-

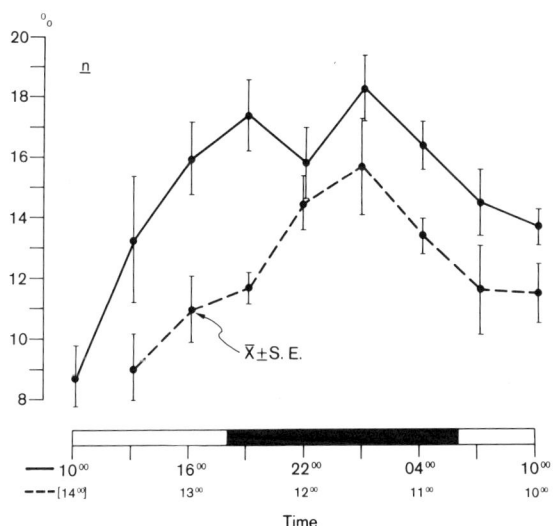

Fig. 2: ——— = LD(12:12)-controls, – – – = ΔɸLD(12:12)-manipulated group. Dark spans (black bar) of lighting schedules are plotted on top of each other; times of sampling (h) at resp. distance of mid-dark phase.

Time	n (%) $\bar{x} \pm$ S.E. (N)	m (%) $\bar{x} \pm$ S.E. (N)
10.00	8.51 ± 0.96 (4)	14.12 ± 1.40 (4)
13.00	13.03 ± 2.11 (4)	21.09 ± 2.96 (4)
16.00	15.77 ± 1.24 (4)	25.17 ± 2.62 (4)
19.00	17.10 ± 1.15 (4)	28.26 ± 1.57 (4)
22.00	15.53 ± 1.22 (4)	24.90 ± 1.51 (4)
01.00	18.02 ± 1.10 (4)	28.54 ± 1.43 (4)
04.00	16.09 ± 0.83 (4)	25.54 ± 2.31 (4)
07.00	14.16 ± 1.11 (4)	22.92 ± 2.08 (4)
10.00	13.43 ± 0.56 (4)	22.07 ± 0.89 (4)
M_{24h}	14.63 ± 0.57 (36)	23.62 ± 0.90 (36)
$\bar{x}_{max}$ (%)	132.2	120.8
$\bar{x}_{min}$ (%)	58.1	59.8
p(F)	<0.01	<0.01
p(t), $\bar{x}_{max}-\bar{x}_{min}$	<0.001	<0.001
p(t), $\bar{x}_{max}-M_{24h}$	n.s.	n.s.
p(t), $\bar{x}_{min}-M_{24h}$	<0.005	<0.005

Table 1: Binucleated liver cells. LD(12:12)-controls.

Time	n (%) $\bar{x} \pm$ S.E. (N)	m (%) $\bar{x} \pm$ S.E. (N)
(10.00)	- - -	- - -
13.00	8.81 ± 1.17 (4)	14.96 ± 2.06 (4)
16.00	10.71 ± 1.10 (4)	17.29 ± 1.53 (4)
19.00	11.35 ± 0.49 (4)	17.50 ± 0.99 (4)
22.00	14.23 ± 0.90 (4)	22.51 ± 1.40 (4)
01.00	15.38 ± 1.58 (4)	22.51 ± 1.00 (4)
04.00	13.14 ± 0.55 (3)	21.27 ± 1.01 (3)
07.00	11.31 ± 1.48 (4)	18.41 ± 1.80 (4)
10.00	11.21 ± 0.95 (4)	17.83 ± 1.84 (4)
M_{24h}	11.98 ± 0.50 (31)	18.96 ± 0.67 (31)
$\bar{x}_{max}$ (%)	128.4	118.7
$\bar{x}_{min}$ (%)	73.5	78.9
p(F)	<0.05	<0.05
p(t), $\bar{x}_{max}-\bar{x}_{min}$	<0.02	<0.025
p(t), $\bar{x}_{max}-M_{24h}$	<0.05	n.s.
p(t), $\bar{x}_{min}-M_{24h}$	n.s.	n.s.

Table 2: Binucleated liver cells. Δϕ LD(12:12)-manipulated animals.

what lower than in the corresponding animals of the LD-reference group. Consequently, the 24h-mean value of $\underline{n}$ under LD (12:12) was also slightly but significantly lower by 2.65% ($p(t) < 0.005$).

$\underline{m}$ (= $\underline{n}$ corrected by Pfuhl's formula): In the LD-standardized group the percentual numbers ($\underline{m}$) of BLC at the different time points, as obtained from Pfuhl's correction, generally increased by a fairly constant factor (~ 1.6) over the respective gross-count values ($\underline{n}$).

When compared to $\underline{n}$ the circadian fluctuation of BLC indicated by $\underline{m}$ were strongly accentuated without showing any remarkable change of their phasing (fig. 3). The difference between $\underline{m}$ at the peak and trough amounted 14.4%. Also, the overall-mean achieved a high level of 23.62% (table 1), as a result of which the relative amplitude decreased by 4.55%.

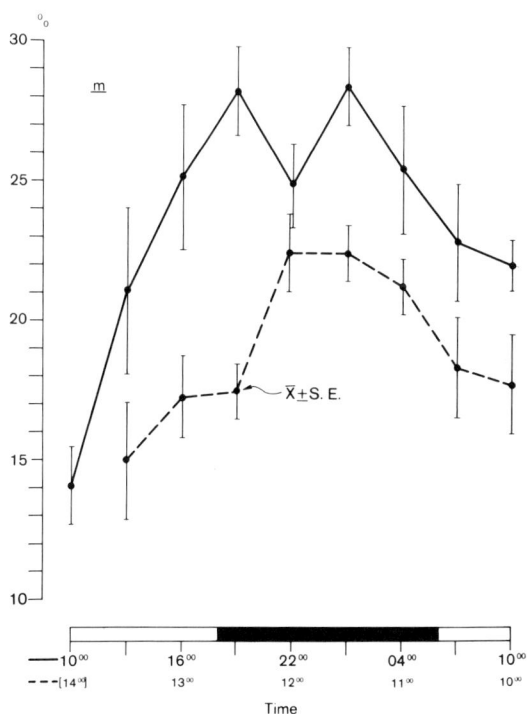

In the LD(12:12)-manipulated animals, the karyometric corrections of the BLC gross counts resulted in a change of their time-related variations similar to that observed in the LD-controls. Here too, the quantitative relation of $\underline{m}$ to $\underline{n}$ equalled about 1.6 (table 2). The 24h-mean level of $\underline{m}$ in this group (18.96%) was significantly lower (4.66%, $p(t) < 0.01$) than that of the controls.

Except for the values obtained at 13.30, the patterns of the corrected BLC counts in both animal

Fig. 3: Same explanation as given for fig. 2.

DISCUSSION

The results obtained from this study clearly demonstrate the existence of a pronounced circadian rhythm in the relative number of binucleate cells in adult rat liver. They confirm and extend similar findings reported in the literature (5,6, 18,29). The adaptive nature of BLC formation is evident from the effective phase-synchronization of its rhythmic variation pattern to the altered light schedule. The influence of the lighting schedule on the phasing of the BLC rhythm is rather indirect; i.e., it mainly acts by synchronizing the animal's sleep-wake and feeding cycle. This assumption is not unlikely, since it is known that the feeding pattern also completely overrules the effect of light on the mitotic activity in the liver (23,27).

The slight depression of the BLC value in the light-manipulated animals is quite similar to that observed in shift studies on other rhythmic variables (22). It may signal

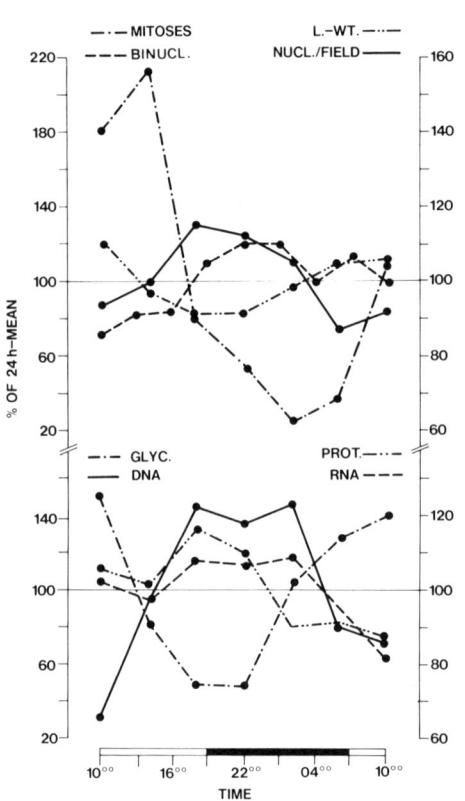

Fig. 4: Phase-relationship of nucleic and cytoplasmic rhythmic activities in rat liver. Mean values are presented as percent values of 24-h overall mean (= 100). (From Abicht J. (1977). Die Circadianrhythmik des DNS-Gehaltes in der Rattenleber. Neue Untersuchungen unter Berücksichtigung der Methodischen Fehlerquellen und der Circadianrhythmen anderer Leberwerte. Doctoral dissertation, Hannover)

the long-term effect of partial rhythm-uncoupling initiated by the sudden time-displacement. This condition is known as disadaptation stress.

The drastic increase and subsequent decrease of BLC rates within the 24-hour time span, in combination with the well-known very low number of hepatocellular mitotic figures in the adult organism favour the hypothesis of Bucher (6), who proposed to explain BLC circadian rhythms by the occurrence of amitosis-and-nuclear fusion cycles. The fundamental meaning of such cycles, which take place during interphase (6,20), is the net gain or reduction, respectively, of the nucleic surface in response to varying demands for specific cell work. In view of this, the dark-phase-associated, high BLC level reflects the nucleic surface enlargement as a reaction to increased metabolic loads during the rat's state of activity and food ingestion. This enlargement is reversed by BLC rate reduction during the circadian resting phase.

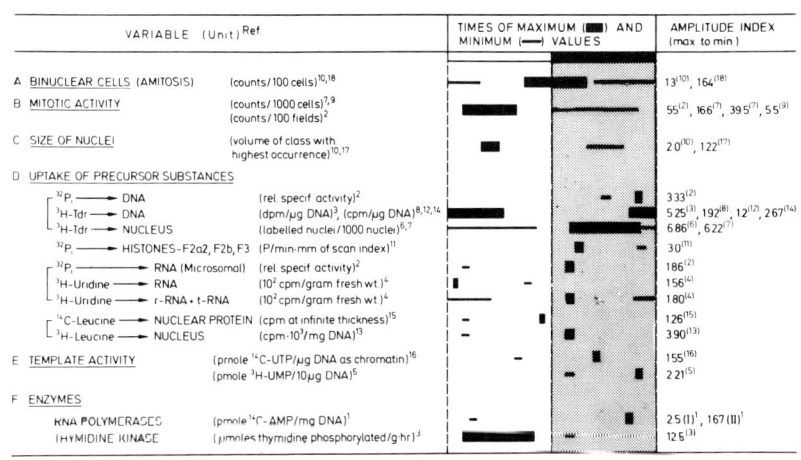

Fig. 5: Circadian rhythmic activities of liver cell nuclei in the rat and mouse. Dark span: ~18.00 - 06.00 (shaded area), light span: ~06.00 - 18.00). Rhythm amplitudes are presented each as the quotient of the circadian max. and min. mean value.

In rat and mouse liver, significant rhythmic variations of DNA concentration have been independently observed by several investigators (1,9,12,21,25). In a previous study it was found that BLC rates increase and decrease along with the DNA concentration of the livers (fig. 4) (21). Basically the same temporal correlation with other nucleic activities is obtained from comparison of independent daytime-qualified investigations reported in the literature (fig. 5). This findings suggests that, along the 24-hour time scale, the alternating stages of BLC formation and reduction take place in the frame of complex functional rearrangements in the hepatocyte nucleus. These have been shown to be ultimately related to qualitative and, probably also, quantitative changes of DNA activity (9,28).

REFERENCES

1. Abicht J, Philippens KMH (1973). Circadian rhythmicity of liver DNA in rats. Int J Chronobiol 1:317.
2. Barbiroli B, Moruzzi MS, Monti MG, Tadolini B (1973). Diurnal rhythmicity of mammalian DNA-dependent RNA polymerase activities I and II: Dependence on food intake. Biochem Biophys Res Comm 54:62.
3. Barnum CP, Jardetzky CD, Halberg F (1958). Time relations among metabolic and morphologic 24-hour changes in mouse liver. Am J Physiol 195:301.
4. Boehm N, Moser B (1976). Reversible Hyperplasie und Hypertrophie der Mäuseleber unter funktioneller Belastung mit Phenobarbital. Beitr Path 157:283.
5. Bucher O (1966). Tagesrhythmiches Verhalten von polyploiden Großkernen, Amitose, zweikernige Zellen und Kernverschmelzung im Leberparenchym der Ratte. Anat Anz 118:452.
6. Bucher O, Suppan P (1967). Amitose et fusion nucléaires au cours du rhythme circadien. In Mayersbach Hv (ed): "The Cellular Aspects of Biorhythms," Berlin: Springer, pp 123.
7. Dallman PR, Spirito RA, Siimes MA (1974). Diurnal patterns of DNA synthesis in the rat: Modification by diet and feeding schedule. J Nutr 104:1234.
8. Döring R, Rensing L (1973). Circadian rhythm of different RNA fractions in rat liver and the effect of cycloheximide. Comp Biochem Physiol B 45:285.
9. Earp HS (1974). Glucocorticoid regulation of transcription: the role of physiologic concentrations of adrenal glucocorticoids in the diurnal variation of rat liver chromatin template availability. Biochem Biophys Acta 340:95.

10. Echave Llanos JM, De Vaccaro MEE, Surur JM (1970). 24-hour variations in DNA of the liver in young and adult male mice. J Interdiscipl Cycle Res 1:161.
11. Echave Llanos JM, Aloisso MD, Souto M, Balduzzi R, Surur JM (1971). Circadian variations of DNA synthesis, mitotic activity and cell size hepatocyte population in young immature male mouse growing liver. Virchows Arch Abt B Zellpath 8:309.
12. Eling W (1967). The circadian rhythm of nucleic acids. In Mayersbach Hv (ed): "The Cellular Aspects of Biorhythms," Berlin: Springer, pp 105.
13. Hardeland R, Hohmann D, Rensing L (1973). The rhythmic organization of rodent liver, a review. J Interdiscipl Cycle Res 4:89.
14. Jackson B (1959). Time-associated variations of mitotic activity in livers of young rats. Anat Rec 134:365.
15. Jerusalem C, Eling W, Yap P (1970). Histochemische und elektronenmikroskopische Veränderungen der Leberzelle im Tagesrhythmus und unter experimentellen Bedingungen. Acta Histochem (Jena) 36:168.
16. Letnansky K (1974). Zirkadiane Rhythmen bei der Phosphorylierung von Kernproteinen und ihre Bedeutung für die Zellproliferation. Wien Klin Wochenschr 86:250.
17. Münzer FT (1924). Experimentelle Studien über die Zweikernigkeit der Leberzellen. Ibid 104:138.
18. Omochi SH, Nagata T, Momoze S (1957). Hourly variation of the frequency of cell division and the fate of binucleate cells in rat liver. Acta Anat Nipponica 32:416.
19. Pfuhl W (1930). Untersuchungen über zweikernige Zellen 1. Mitteilung. Die Berechnung der zweikernigen Zellen nach der Auszählung im mikroskopischen Schnitt. Mikr anat Forsch 22:557.
20. Pfuhl W (1938). Die mitotischen Teilungen der Leberzellen im Zusammenhang mit den allgemeinen Fragen über Mitose und Amitose. Anat Entwicklungsgesch 109:99
21. Philippens KMH, Abicht J (1975). Tagesrhythmik des Nukleinsäurestoffwechsels. Nova Acta Leopoldina 46(225):143.
22. Philippens KMH (1976). The manipulation of circadian rhythms. Arch Toxicol 36:277.
23. Philippens KMH (in press). Synchronization of rhythms to mealtiming. NATO Adv Study Inst on Princ Appl Chronobiol, Hannover (1979).
24. Potter VR, Gebert RA, Pitot HC, Peraino C, Lamar Cjr, Lesher S, Morris HP (1966). Systematic oscillations in metabolic activity in rat liver and in hepatomas. I. Morris Hepatoma No. 7793. Cancer Research 26:1547.

25. Ruby JR, Scheving LE, Gray SB, White K (1973). Circadian rhythm of nuclear DNA in adult rat liver. Exper Cell Res 76:136.
26. Sestan N (1964). Diurnal variations of ^{14}C-leucine incorporation into proteins of isolated rat liver nuclei. Naturwissenschaften 51:371.
27. Schulte-Hermann R, Landgraf H (1974). Circadian rhythm of cell proliferation in rat liver: synchronization by feeding habits. Z Naturforsch CC 29:421.
28. Steinhardt WL (1971). Diurnal rhythmicity in template activity of mouse liver chromatin. Biochim Biophys Acta 223:301.
29. Zhirnova AA (1969). Relationships between diurnal rhythm in number of binuclear cells in rat liver and its glycogen-forming function. B Eksp Biol Med 68:98.

FLUCTUATIONS IN NUCLEAR AND CYTOPLASMIC SIZE OF VAGINAL AND
BUCCAL EPITHELIAL CELLS REFLECT THE TIME OF THE OVULATION AS
WELL AS THE TIME OF THE DAY

W.J. Rietveld and M.E. Boon

Department of Physiology, University of Leiden,
The Netherlands; SSDZ, Department of Pathology,
Delft, The Netherlands.

In recent years a considerable amount of data has been acquired about 24 h or circadian fluctuations in the ultra morphology as well as the chemical constituents of the living cell and of its different compartments (v. Mayersbach 1978; Muller 1971). The intracellular location of mitochondria, endoplasmic reticulum, the lysosomes and the Golgi apparatus exhibit periodic reorganization. Quantitative changes of glycogen as well as granules and vesicles containing hormones and neurotransmitters occur, and even the DNA content is influenced by a 24 h rhythm (Scheving et al. 1974). Exfoliating cells of squamous epithelium are known to be under hormonal influence. For example, the degree of maturation of the squamous epithelium of the female genital tract is hormone dependent. Ovarian steroids such as estrogen and progesterone are able to activate the DNA-nucleoprotein complex, forming RNA which in its turn stimulates the synthesis of various enzymes and structural proteins.
The maturing ovarian follicle promotes growth and maturation of the squamous cells up to and including the superficial layer. The small parabasal cells ripen via large parabasal and intermediate cells to precornified and cornified ones. After the ovulation on day 14 in a 28 day cycle, the follicle is transformed into a corpus luteum which produces progesterone for the next 14 days causing a rapid desquamation of the topmost layer as well as a decrease in the degree of maturation.
In an attempt to confer numerical reproducibility upon hormonal evaluation, various indices have been proposed (Koss 1979; Pundel 1957). Many of these are based on the maturation effect of the steroid hormones on the squamous cells expres-

sing, for example, the percentage of superficial squamous cells with pycnotic nuclei to all mature squamous cells (the karyopycnotic index). The maturation index expresses the maturation of the cells as the percentage of parabasal cells to intermediate to superficial cells. The use of these indices in the study of vaginal, cervical, and buccal cells during the menstrual cycle reveals a sharp increase of up to 70% of the basal value at the time of the ovulation (Boon 1980).

In an attempt to provide objective parameters, and to obtain more insight into this phenomenon, we made a more detailed study of the changes in the squamous cell population during the ovulation time by measuring cell as well as nuclear area. We included both vaginal and buccal cell samples. However, as the DNA content of the individual cells is reported to fluctuate across the day, one may put forward the question of how far this fluctuation will be reflected in some morphological aspects as well, superimposed upon or interacting with hormonal controlled changes. We therefore included a study on the effect of the time of the day on the buccal cells as well. Vaginal and buccal smears were prepared with an Ayre spatula. The smears were air dried and stained with the May-Grünwald-Giemsa method. We choose the air-dry May-Grünwald-Giemsa method for two reasons. First, in the otherwise often used wet-fix Papanicolaou method, the cells are much smaller due to shrinkage (Beyer-Boon et al. 1979); and we anticipated that subtle variations in size would be difficult to assess with that method. Second, since no fixative is needed at the moment the smear is made the May-Grünwald-Giemsa method was also more convenient.

Two sterilized woman of reproductive age who were known to have ovulatory cycles took a buccal and vaginal sample at day 10, 11, 13, 14, 15, 17, 18, 19, 22 and 25 at 0700 h. Basal temperature measurement was done to evaluate the time of ovulation. From each slide, 25 cells were randomly selected. The nuclei and cytoplasm of the cells were measured with a Leitz ASM image analysis (100 x oil immersion). The ASM image analysis system was equipped with a camera lucida system, and the digital cursor was seen in the microscopic field under study. Thus, the selected object could be outlined. The computer calculated from the two delineated structures the following 3 cell features: nuclear area, cell area and nuclear cytoplasmic (N/C) area ratio. From each case the mean and the standard deviation of the 3 listed cell features were calculated: this resulted in 6 parameters per case.

To exclude variations caused by the menstrual rhythm in the study on the effect of the time of the day, another study was conducted with male subjects only. During 5 consecutive days, two buccal smears were taken at 3 hour intervals throughout 24 hours using the same method as described above. After staining, nuclear measurements only were done in each slide on 25 selected intermediate squamous cells showing no signs of pycnosis, bacterial contamination or cytolysis. After analysis of the data obtained from one subject during the five days, repeated measurements were done on three other subjects at 0200 and 1400 hours.
The data from 5 consecutive days were averaged on a 24-h scale. Using the single cosinor procedure (Halberg et al. 1972), a least-squares fit of a cosine model was done with a Texas Instruments TI 59 pocket calculator. In this way an estimation of the mean value (mesor), double amplitude and acrophase with their respective standard errors in addition to the associated p-values obtained in the zero amplitude test were given.

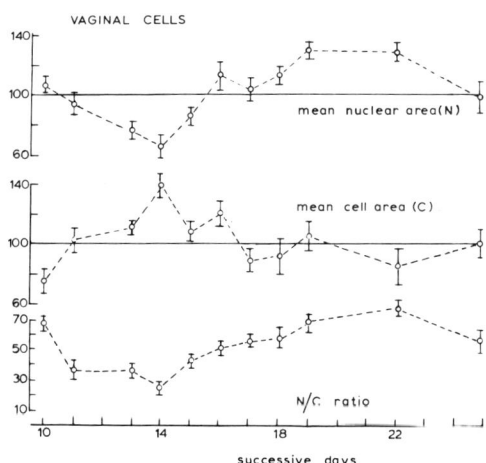

Fig. 1. Top: mean nuclear area ($\pm$ 1 S.E.M.) of vaginal cells expressed as percentage of the average value. Middle: The same for the cell area. Bottom: The nuclear/cell ratio multiplied by 10^3.

Fig. 1 shows the results obtained from the vaginal cells from one woman. Both nuclear and cell size is expressed as percentage (mean $\pm$ S.E.M.) of the average value during the period measured. At Day 14 (the time at which the ovulation took place according to the basal temperature measurement), the nuclear area (upper graph) decreased, whereas at the same time the cell size (middle graph) increased. The percentage increase in cell area exceeded the percentage decrease in nuclear size as is pictured in the N/C ratio (lower graph). The data from the buccal cell measurements are presented in Fig. 2.

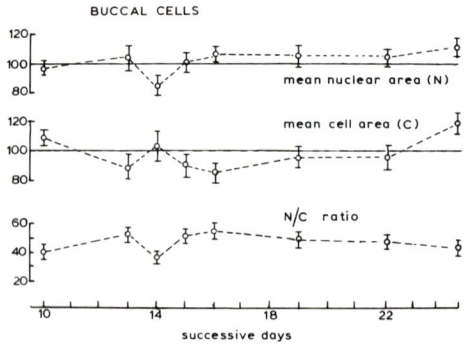

Fig. 2. Buccal cell measurements. For legend see Fig. 1.

Although the percentage change at Day 14 was less marked than in the case of the vaginal cells, these cells also showed a substantial decrease in nuclear area accompanied by a (larger) increase in cell area.
Statistical analysis of the measured values at Day 14 as compared with the values at Days 10, 13, 15 and 19 (a priori t-test, $0.01 < p < 0.05$) revealed a significant difference in all cases (vaginal as well as buccal cell and nuclear area). Similar findings were obtained from the data of the second woman.
The results from the second experiment in which the average nuclear area was measured at the 8 different time points is presented in figure 3.

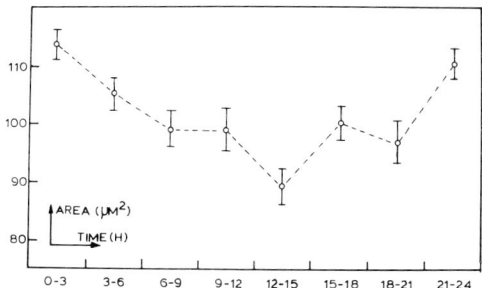

Fig. 3. Plot of the nuclear area at different times of the day.

Nuclear size appeared to be larger during night-time hours while minimum value was reached between 1200 and 1500 hours. Using a cosine function as a model to describe the rhythm paramaters, it was found that 76% of the total variance could explained by assuming a 24-h cosine rhythm with a p value of < 0.02. The mean area (M) was 102.12 ($\pm$ 1.6) μm^2 with a total deflection (DA) of 18.0 ($\pm$ 4.5) μm^2. The calculated acrophase (ϕ = maximum value in respect to **2400 h)** was 2.41 ($\pm$ 0.57) h. The data of the three other subjects measured at 0200 and 1400 h respectively gave similar results: 104.2 $\pm$ 2.7; 99.1 $\pm$ 2.1; 105.3 $\pm$ 3.2 measured at 0200 h as compared to 97.1 $\pm$ 2.5; 90.3 $\pm$ 1.9 and 100.1 $\pm$ 2.8 at 1400 h. In all cases the value of 0200 h exceeded the value of 1400 h.

In discussing the results of both experiments, two possible explanations for the observed changes in nuclear and cell area during the time of ovulation can be given.

First, the measured decrease in nuclear area may reflect an increase in the number of matured intermediate cells during the time of estrous which is preceded and followed by a decrease. On the other hand, temporary differences in function and/or biochemical structure of the cytoplasm per se may be involved. For example since the binding of the steroid hormones to the cytomembrane leads to activation of mRNA followed by the synthesis of enzymes and structural proteins, biochemical alterations induced by the hormonal changes at ovulation may be responsible for both the differences in cell and nuclear area. Which one of these two hypotheses is responsible for the observed changes will be a

matter for future investigation. The first question to be
answered in the second experiment, is whether these changes
in nuclear area are endogenous to the peripheral cell or induced by some exogenous factor (e.g. temperature, hormone
concentration, meal timing). In this respect, it is note-
worthy that by recording the number of mitoses in the human
epidermis, a similar 24-h pattern has been described (Scheving
et al. 1974) with a peak at the early night time hours. If
the nuclear area would be under hormonal control as it is the
case in the female, one would expect some effect of fluctuat-
ing androgens. Circadian fluctuations in plasma testosterone
have been described (Lincoln et al. 1974) with a maximum
value at about 0700 h and a minimum at 2200 h. Thus a time
lag of about 5 hours should be assumed between this rhythm
and the variations in nuclear size. More experiments are
needed to provide information about the possible endogenous
character of the described nuclear changes, and experiments
under more controlled conditions are in progress.
Summarizing, it can be said that the estrous cycle as well as
the time of the day are reflected in the morphology of the
peripheral cells and that biochemical experiments are needed
to obtain more insight in the nature of the observed changes.

References

Beyer-Boon ME, et al. (1979). Effect of various routine cyto-
  preparatory techniques on normal urothelial cells and
  their nuclei. Acta Path Microbiol Scand 87: 63.
Boon ME, Tabbers-Boumeester M (1980)"Gynacological Cytology".
  A textbook and atlas. The MacMillan Press Ltd., London,p 46.
Halberg F, et al. (1972). Autorhythmometry procedures for
  physiologic selfmeasurements and their analysis. Physio-
  logy Teacher 1: 1.
Koss LG, (1979). "Diagnostic Cytology". J.B. Lippincott Com-
  pany, Philadelphia, Toronto p 203.
Lincoln GA, et al. (1974). The circadian rhythm in plasma
  testosterone concentration in man. In "Chronobiological
  aspects of Endocrinology". F.K.Schattauer Verl. Stuttgart.p.13
Von Mayersbach H, (1978). Die Zeitstruktur des Organismus.
  Arzneim.-Forsch./Drug Res.28 (II): 1850.
Müller O, (1971). Acta Histochem. Supl. X: 141.
Pundel JP, (1957)"Acquisitions récentes en cytologie vaginale
  normonale". Masson, Paris p 143
Scheving LE, Pauly JE, (1974). Circadian rhythms, clinical
  application. Chronobiologia 1: 3.

CHRONOHISTOCHEMISTRY

CIRCADIAN CHANGES OF LYSOSOMAL ENZYME ACTIVITIES IN RAT
HEPATOCYTES USING ULTRACYTOCHEMISTRY

Yasuo Uchiyama and Heinz von Mayersbach*

Department of Anatomy, Tohoku University School
of Medicine, Sendai 980, Japan
*Department Anatomie, Medizinische Hochschule
Hannover, D-3000 Hannover 61, West Germany

INTRODUCTION

Lysosomes are subdivided into primary and secondary lysosomes and are commonly refered to as "the lysosomal system" (Schllens et al., 1977). Whether the enzyme activities in lysosomes are heterogeneous or homogeneous has been controversial. Some researchers have studied them biochemically (Pertoft and Wärmegard, 1978), and others have studied them histochemically (Nichols et al., 1971). Chronobiological approaches also have contributed to the understanding of this question and supported the concept of heterogeneity in enzyme activities of lysosomes; biochemical and histochemical techniques at the light and electron microscopic levels have been made using the chronobiological approach (Bhattacharya, 1977; Uchiyama and von Mayersbach, 1979; Uchiyama et al., 1980).

In order to further understand lysosomes, the authors have investigated circadian variations in different lysosomal enzymes; these have included acid phosphatase, arylsulphatase and ß-glucuronidase of rat hepatocytes during a 24-h span using ultracytochemistry, and have shown marked changes in incidence of lysosomes showing each enzyme activity.

MATERIALS AND METHODS

Thirty male adult Wistar rats (AF/Han, ca. 300g) were employed in these studies. The animals were subdivided into 6 subgroups which were housed in plastic cages and kept in ventilated chambers with relatively constant environmental conditions: temperature, 22 ± 1°C; relative humidity, 55%, an artificial 12:12 h light-dark cycle (light period: 0900-2100); and they were fed ad libitum.

Sampling began at 1300 on April 3, 1978, and was continued every 4h for 24h. At each sampling 5 rats from each rat group were killed by rapid decapitation. Liver samples were immediately excised from rats and cut into small blocks (1 x 1 x 1 mm). For enzyme histochemical reactions of acid phosphatase and arylsulphatase, the specimens were fixed in 2% glutaraldehyde buffered with 0.1M cacodylate-HCl buffer, pH 7.3, containing 7.5% sucrose for 2h. For the reaction of ß-glucuronidase, they were fixed in 4% formalin sucrose for 4h. After washing for 12h, 40μm sections were prepared with a tissue sectioner (Sorvall TC-2).

Histochemistry: The liver tissue was treated with the following enzyme reaction media: 1) The incubation for acid phosphatase was performed after Gomori (1950) (temperature: on ice, time: 20 min.), 2) arylsulphatase after Hapsu-Havu et al. (1967) (temperature: 37°C, time: 45 min.) and 3) ß-glucuronidase after Hayashi et al. (1968) (instead of alcohol pretreatment, the incubation time was prolonged. temperature: 37°C, time: 60 min.).

Then, specimens were postfixed with 1% OsO_4 buffered with 0.1M cacodylate-HCl, pH 7.3, containing 7.5% sucrose for 1h. They were dehydrated with graded alcohols and embedded in Epon 812. Silver sections were cut by Reichert OmU2 and Porter-Blum MT-2 ultramicrotome and observed by Zeiss EM9S and Hitachi H-600 electron microscopes without staining.

Measurement of lysosomes displaying each enzyme activity: Fifty hepatocytes in peripheral parts of liver lobules, chosen randomly per rat at each time-point, were photographed at a magnification of 1900x. After printing at 5 times the original magnification, lysosomes showing each enzyme activity per intersected hepatocyte were counted.

RESULTS AND DISCUSSION

As shown in Fig. 1, the incidence of lysosomes exhibiting acid phosphatase, arylsulphatase and ß-glucuronidase activities respectively per intersected hepatocyte demonstrated a statistically significant circadian rhythm, and the phasing of these marker-enzyme activities was not the same inasmuch as each of these marker enzymes displayed maximal and minimal activities at different circadian stages (clock hours).

1) Acid phosphatase: Peaks in the incidence of lysosomes exhibiting acid phosphatase activity per intersected hepatocyte occurred at 1300 and 0100, and troughs at 0900 and 2100. The activity was predominantly localized in lysosomes located near bile canaliculi at 1300 (Fig. 2a). At the other time points, however, it was also found in ground plasm, especially at 0100 (Fig. 2b). At 0100, lysosomes showing the activity were relatively voluminous and peculiar in form. As to the ground plasm activity, it remains an open question whether it originates from lysosomes or from other cytoplasmic organelles, or is it really localized. In

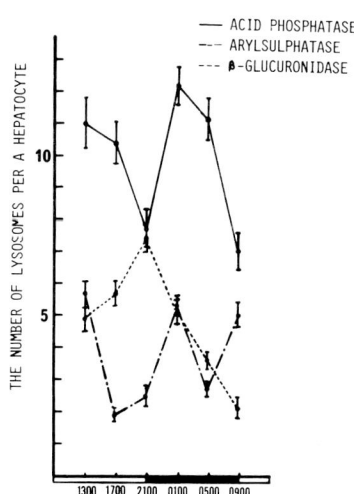

Fig. 1. Changes in incidences of lysosomes exhibiting acid phosphatase, arylsulphatase and ß-glucuronidase activities respectively per intersected hepatocyte during a 24-h span. Standard errors are marked on each bar ($p < 0.01$).

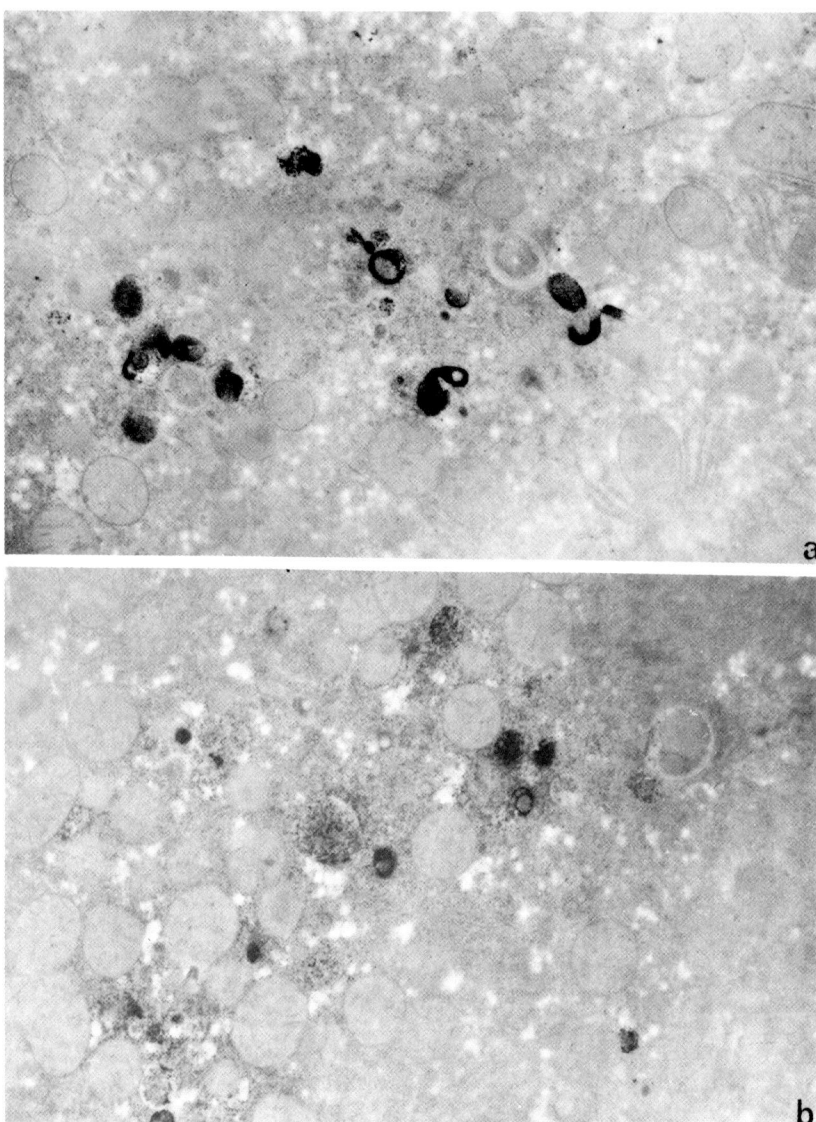

Fig. 2. Parts of hepatocytes after the acid phosphatase reaction. a. At 1300h. Lysosomes showing a acid phosphatase activity are clearly demonstrated. b. At 1700h. The activity is seen not only in lysosomes, but in ground plasm. x 10000.

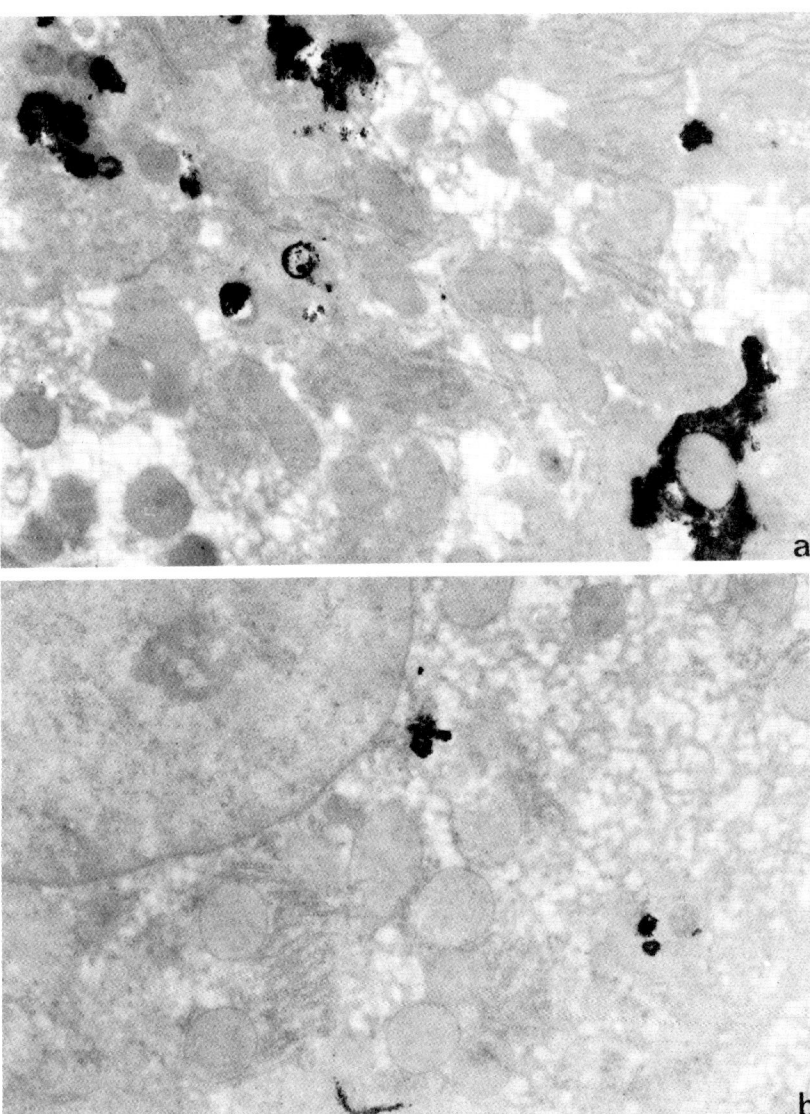

Fig. 3. Parts of hepatocytes after the arylsulphatase reaction. a. At 1300h. The activity is demonstrated in lysosomes and patches of the endoplasmic reticulum. b. At 1700h. Lysosomes exhibiting the activity are rarely seen. However, it is clearly demonstrated in the cistern of the Golgi lamellae. x 11500.

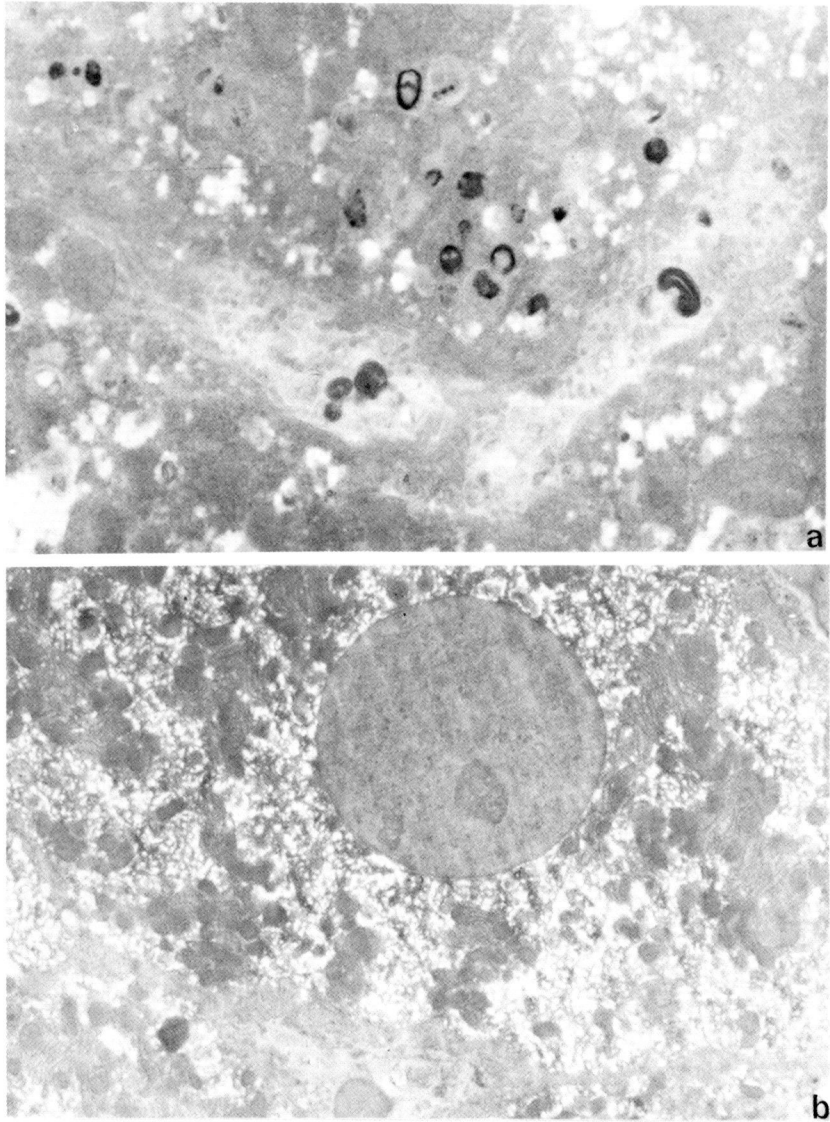

Fig. 4. Parts of hepatocytes after the ß-glucuronidase reaction. a. At 2100h. Numerous lysosomes displaying the activity are gathered near bile canaliculi. x 10000. b. At 0900h. No lysosomes showing the activity can be seen in the cytoplasm. x 4600.

addition, although it had been demonstrated in the Golgi apparatus and endoplasmic reticulum, the activity within them was not enough to make meaningful comparisons.

2) <u>Arylsulphatase:</u> The incidence of lysosomes showing a arylsulphatase activity per intersected hepatocyte was high at 1300 and 0100 and low at 0500 and 1700. The activity discerned in lysosomes was located not only near bile canaliculi, but also was found throughout the cytoplasm as well as in patches of the endoplasmic reticulum (Fig. 3a). Moreover, the activity was also clearly localized in the cistern of the Golgi lamellae at 0500 and 1700, a time when the activity in lysosomes and patches of endoplasmic reticulum was considerably low (Fig. 3b). This suggests the cytoplasmic pathway of arylsulphatase in hepatocytes during the 24-h span.

3) <u>ß-glucuronidase:</u> The peak of the incidence of lysosomes displaying a ß-glucuronidase activity per intersected hepatocyte occurred at 2100, and their number decreased during the dark cycle. The trough occurred at 0900, and the number of lysosomes showing the activity increased during the light cycle. The activity was localized in lysosomes located near bile canaliculi (Figs. 4a,b), although occasionally it also was found in fat-droplet-like bodies. The activity in these organelles was completely inhibited by glucosacchaxo-1,4-lactone.

Judging from the fact that the pattern of activity for each enzyme studied in lysosomes is clearly independent from each other, lysosomes in rat hepatocytes appear to be definitely heterogeneous with respect to their regulation and/or possibly also their distribution in the different subpopulations within the lysosome compartment. Moreover, this may mean that lysosomal subpopulations are derived from different cytogenic pathways as described by Dvořák (1974); and/or that lysosomes at the beginning from the Golgi apparatus or endoplasmic reticulum include a single enzyme or a very small number of enzymes, and after repeated fusion they become mature primary lysosomes.

REFERENCES

Bhattacharya R (1977). Circadian differences of lysosomes and their enzymes. Nova Acta Leopoldina 46:171.

Dvořák M (1974). Origin and development of lysosomes and peroxisomes. In Dvořák M (ed): "Biogenesis of cell organelles," Brno: Purkyne University, p 59.

Gomori G (1950). An improved histochemical technique for acid phosphatase. Stain Tech 25:81.

Hapsu-Havu VK, Arstilla AU, Helminen HJ, Kalimo HO (1967). Improvements in the method for electron microscopic localization of arylsulphatase activity. Histochemie 8:54.

Hayashi M, Shirahama T, Cohen AS (1968). Combined cytochemical and electron microscopic demonstration of ß-glucuronidase activity in rat liver with the use of a simultaneous coupling azo dye technique. J Cell Biol 36:289.

Nichols BA, Bainton DF, Farquhar MG (1971). Differentiation of monocytes. Origin, nature and fate of their azurophil granules. J Cell Biol 50:498.

Pertoft H, Wärmegard B (1978). Heterogeneity of lysosomes originating from rat liver parenchymal cells. Biochem J 174:309.

Schellens JPM, Daems WTH, Emeis JJ, Brederoo P, Bruijn WC, Wisse E (1977). Electron microscopical identification of lysosomes. In Dingle JT (ed): "Lysosomes," Amsterdam-New York-Oxford: North Holland, p 147.

Uchiyama Y, Mayersbach H v (1979). Study on the rhythmic changes of lysosomal enzyme activities in rat liver parenchymal cells using ultracytochemistry. Chronobiologia 2:166.

Uchiyama Y, Groh V, Mayersbach H v. Circadian variations of liver lysosomes. In contribution.

LYSOSOMAL HISTOCHEMISTRY IN RELATION TO A SYNCHRONIZER

Radium Dalwadi Bhattacharya

Department of Physiology, B. J. Medical College,
Ahmedabad 380 016, India

INTRODUCTION

Chronobiological study at the morphological level is a problem that has been addressed by only a small group of investigators. With microscopy, chronobiological studies can be done with:

I. Routine Histology:
Chronobiological variation is seen in nuclear activities like size, number, appearance, mitotic counts, etc.

II. Histochemistry:
This is concerned with the specific recognition and visualization of substrates, e.g., nucleic acid, glycogen, mucosubstances and lipids. It also includes the quantitative evaluation of different enzymes, antigens, antibodies, etc. Histochemistry can be studied at the:
a. Histotopographical level, i.e., in different organ compartments like distribution within the liver lobules, and at the
b. Cytotopographic level, i.e., cellular compartments like Golgi, lysosomes, endoplasmic reticulum, mitochondria, bile canaliculi, etc.

III. Ultrastructural Techniques:
This is the study of cellular constituents and intercellular connections by electron microscopy.

IV. Ultrahistochemistry:
This involves an analysis of substrate, enzymes, etc., at the electron-microscopic level.

In order to prepare tissues for microscopic examination, it is necessary to employ techniques that artificially alter their in vivo state. However, if these techniques are carefully standardized (kept "constant"), any changes seen may be assumed to reflect normal chronobiological variation or the results of experimental conditions imposed on the organism. This is summarized in the following table:

Technique	Biological State	Morphological Appearance, Staining or Histochemical Characteristics
Standardized	Constant	Constant - defined chronobiological state
Standardized	Variable	Variable - different chronobiological state
Variable	Constant	Variable - of restricted use in chronobiological evaluation
Variable	Variable	Potential variability - no evaluation possible

Therefore, to study the chronobiology at the morphological level, the following requisites are necessary:
A. Animal maintenance and handling:
 1. Standardized environment, constant temperature and humidity; controlled light-dark cycle and food.
 2. Avoid stressful conditions in housing, social interactions, transportation and repeat sampling from the same animal.
B. Tissue sampling:
 Work fast to avoid postmortal changes, such as swelling of mitochondria, enzyme activation and inactivation, etc.
C. Tissue preparation:
 1. Carefully controlled fixation, dehydration and embedding.
 2. Carefully controlled in situ perfusion and fixation.
D. Staining and histochemical reactions:
 1. Standardize time and temperature of staining-reaction medium.
 2. Use identical batches of histochemical reagents.
 3. Use fresh chemicals.

The first report of chronobiological variations at the morphological level was by H. von Mayersbach (1967). He showed very distinct 24-hour changes in esterase and succinic dehydrogenase activity as well as in glycogen content of the liver. Bhattacharya (1969, 1978) and Bhattacharya and Wegmann (1971, 1972, 1980) in a series of studies of rat liver, skeletal muscle and cardiac muscle showed a definite 24-hour rhythm in the activity of these enzymes. These studies were carried out in normal environmental conditions with an <u>ad libitum</u> supply of food and water. Later Bhattacharya (1977a, b, 1976, 1980) and Bhattacharya and Mayersbach (1975, 1976, 1979, 1980) reported a circadian rhythm in lysosomes and their enzymes in animals kept under highly standardized conditions. These studies showed that liver lysosomes change their size, number and distribution pattern. They also show different color intensities at different hours of the day, manifesting a rhythm of the enzymes. Uchiyama and Mayersbach (1979) confirmed some of these circadian rhythms by means of electron microscopy.

The circadian rhythm of any variable depends on its "Biodisposability" which is a reference that:
1. may be used to compare the amount of active substance in organs,
2. may be used to determine the dosage per kg of body weight,
3. takes into account the allometric organ/body weight ratio of different sized animals of different ages, and
4. avoids referring to values which are themselves subject to chronobiological changes, such as protein, DNA, etc.

The biodisposability of any variable will be constant in strictly standardized conditions.

Histological, histochemical and ultrastructural studies have demonstrated that chronobiological oscillations are present in cells and their different organelles. The topic considered in this paper is the effect of a synchronizer on these rhythms. Most of the variables that have been studied in rodents manifest a circadian rhythm. Light has been found to be the best synchronizer of these rhythms, i.e., the rodents will synchronize to the light-dark cycle of nature or the laboratory. Generally, if one reverses this light-dark cycle $180°$, the circadian system gradually will invert; but different variables phase shift at different rates. There have been no reports in the literature describing phase shifting of morphological rhythms by reversing the light-dark cycle, except those relating to mitosis. This paper describes the temporal oscillation of liver lysosomes and their enzymes studied in rats after inverting the light-dark schedule 180°.

MATERIALS AND METHODS

Animals. Female Wistar-strain rats weighing 120 gm were used for the experiment. They were housed in groups of four in plastic cages (Mikrolon III, floor space 27x42 cm) in two special climate rooms which were maintained at 22°C ± 1°C with a relative humidity of 55%. Food and water were supplied ad libitum. One room was artificially illuminated (1600 lux fluorescent light) for 12 hours (0600 to 1800) and completely darkened from 1800 to 0600 (LD). The second room was illuminated from 1800 to 0600 followed by 12 hours of complete darkness (DL). Throughout this article, the groups will be referred to as LD and DL rats. The animals were acclimatized to the LD schedule for 4 weeks, followed by DL conditioning for 6 days.

Sampling: Every three hours during a 24-hour period, subgroups of four rats from each group were killed by rapid decapitation, within two minutes after picking up the cage from the climate chamber. Using an assembly-line technique, with 6 teams of 2 persons each, it was possible to weigh, rapidly kill, and perform those procedures necessary to initiate several biochemical and histochemical analyses.

Histochemical techniques. A piece of liver from each animal was removed and fixed in 5% formalin containing 0.44M sucrose at 4°C. After four hours of fixation, the tissues were thoroughly rinsed in 0.44M sucrose and preserved in that solution for 72 hours prior to sectioning. Cryotome sections of 6 μm thickness were prepared, and 2 to 3 sections were mounted on cover slips. The sections were dried for 20 minutes at room temperature, and the histochemical reactions were performed. The acid phosphatase reaction was carried out according to the method described by Barka and Anderson (1963), the B-glucuronidase according to the method of Hayashi (1964), and esterase by the method of Holt (1958).

Semiquantitative analysis. Vitatron was used to analyze the color intensity developed as a result of each enzymatic reaction by using a proper filter for each enzyme reaction product. The different filters used were as follows:

acid phosphatase - filter number 411
B-glucuronidase - filter number 511
esterase - filter number 380

RESULTS

Histochemistry. A histochemical evaluation of lysosomal enzymes has been expressed in arbitrary units and presented in Table 1.

Table 1. Histochemical Evaluation of Enzymes in LD and DL Conditions

TIME	ACID PHOSHATASE		B-GLUCURONIDASE		ESTERASE PH 7.4	
	LD	DL	LD	DL	LD	DL
07	+++±	+++	+++±	+ B	+++	++
10	++	+++	++	++ B	++++	+++
13	++++	++	+± B	++++	+++	++
16	+++	+±	+	+++	++	++
19	+++	+	++++	++±	+++	++
22	++	B	++	++	++	D
01	+±	+	+++	++	+±	+±
04	++ B	++++	+++	+±	D	++

B Big
D droplets

Acid phosphatase. The peak activity of acid phosphatase in LD and DL rats occurred at 1300 and 0400 hours, respectively; whereas the troughs occurred at 0100 in LD and from 1900 to 0100 hours in DL (Fig. 1a). The greatest concentration of the enzyme in the lysosomes advanced 9 hours in the DL conditions (Table 1). It can be seen that the morphological distribution of acid phosphatase is not uniform throughout the 24-hour period. The enzyme activity occasionally is seen as fine dots, big dots or as droplets. At 0400 and 2200 hours, big granules were observed in LD and DL rats, respectively (Fig. 1b).

B-glucuronidase. Table 1 shows peak activity of B-glucuronidase at 1900 hours in LD and 1300 hours in DL rats.

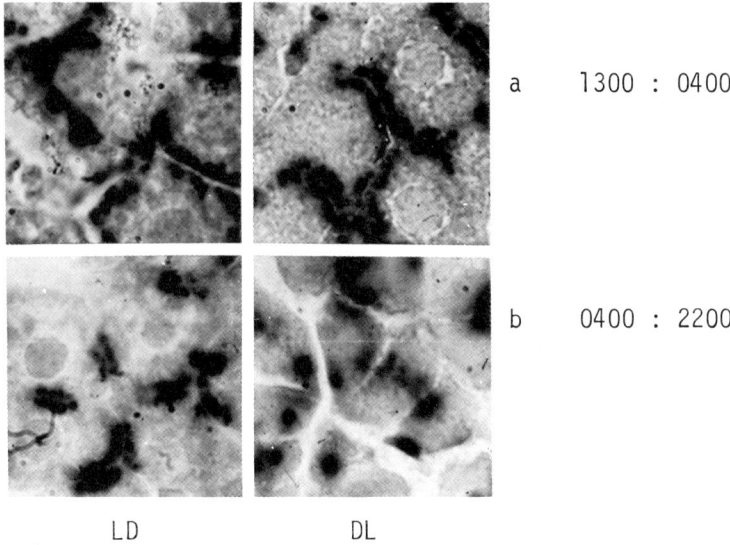

a	1300 : 0400
b	0400 : 2200

LD　　　　　　DL

Figure 1. Acid phosphatase

Minimum activity was observed in LD at 1600 and at 0700 in DL. Figure 2a and b shows the distribution of B-glucuronidase in liver cells at peak and trough times in LD and DL conditions. In the DL condition, the peak advanced by 6 hours and trough by 9 hours. Moreover it can be seen in Figure 2c that the distribution patterns of microsomal and lysosomal B-glucuronidase are different at different hours of the day; this also can be affected by the change in the light-dark schedule. In LD this different microsomal and lysosomal distribution occurred at 1000, whereas it occurred at 2200 hours in DL.

Esterase at pH 7.4. Table 1 shows lysosomal esterase activity at pH 7.4. The peak activity for both LD- and DL-conditioned rats occurred at 1000, and the trough extended from 2200 to 0400 in LD and from 1600 to 0400 in DL rats. Thus at the morphological level the LD:DL conditioning had no effect as far as esterase at pH 7.4 is concerned. Figure 3 shows the distribution of esterase at pH 7.4 as revealed by histochemistry. Although the times of peak and trough did not change, there was definitely a reduction in enzyme activity in the DL condition (Fig. 3a, b). Figure 3c shows

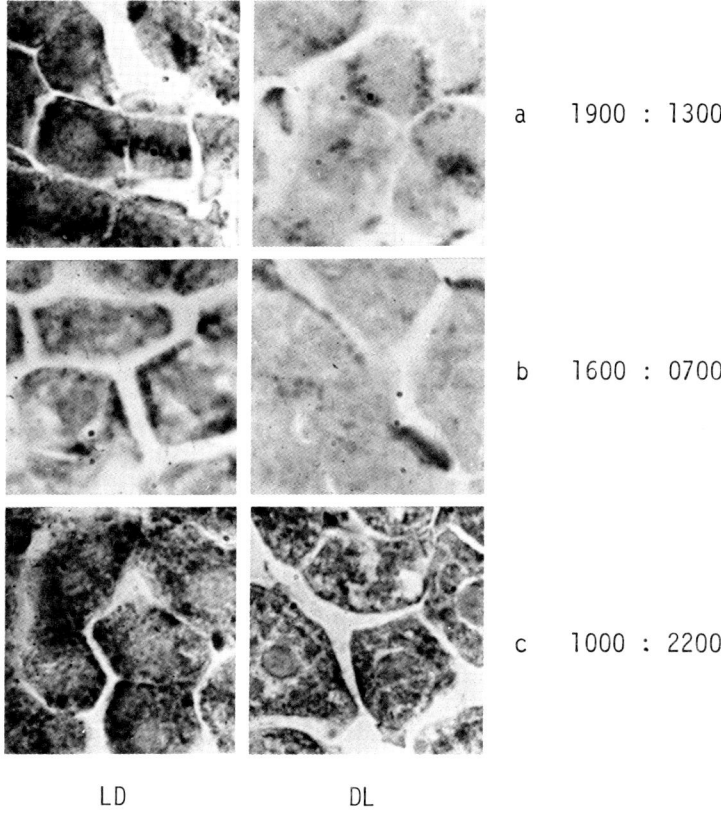

a	1900 : 1300
b	1600 : 0700
c	1000 : 2200

LD   DL

Figure 2. B-glucuronidase.

a big droplet-like structure in liver cells which appeared at 0400 hours in LD rats and at 2000 hours in DL rats.

Semiquantitative Evaluation

Acid phosphatase. Acid phosphatase shows a circadian rhythm in both LD and DL (Fig. 4). The 24-hour mean of enzyme activity was less in DL than in LD, and the difference was statistically significant.

In the LD schedule, peak activity was observed at 1300

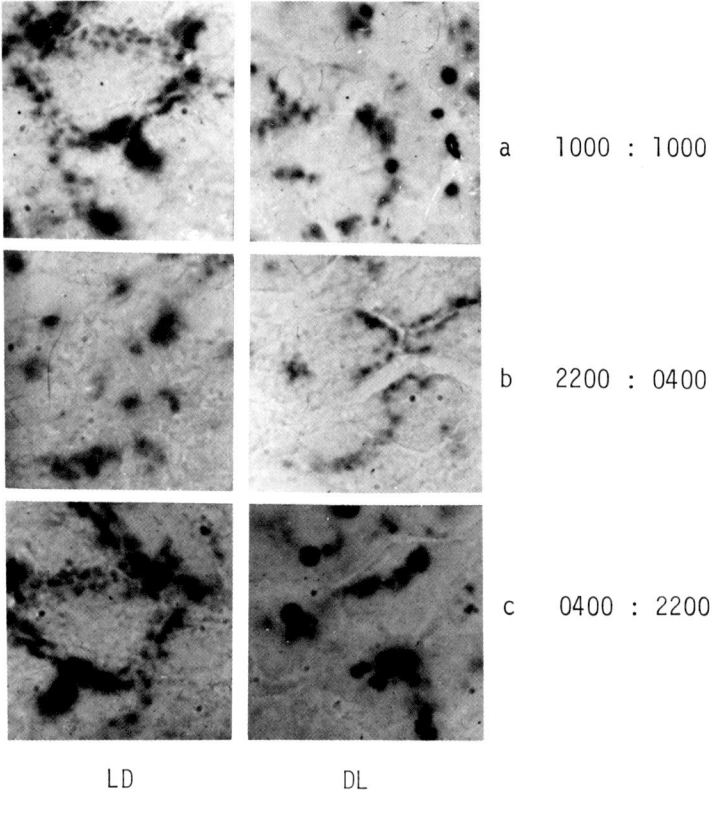

Figure 3. Esterase

hours, and the trough occurred at 2200 hours. Enzyme activity oscillated below the 24-hour mean from 1600 hours to 2200 hours, and it was above the mean level throughout the other time points observed.

In the DL schedule, i.e., when the light was inverted 180°, the enzyme oscillated at a lower amplitude throughout the day. The peak occurred at 0400 hours and the trough at 2200 hours. Thus the peak activity shifted to the left by 9 hours, whereas the trough remained at the same time. In general, the pattern of oscillation in the DL condition is smoother than in LD (Fig. 4).

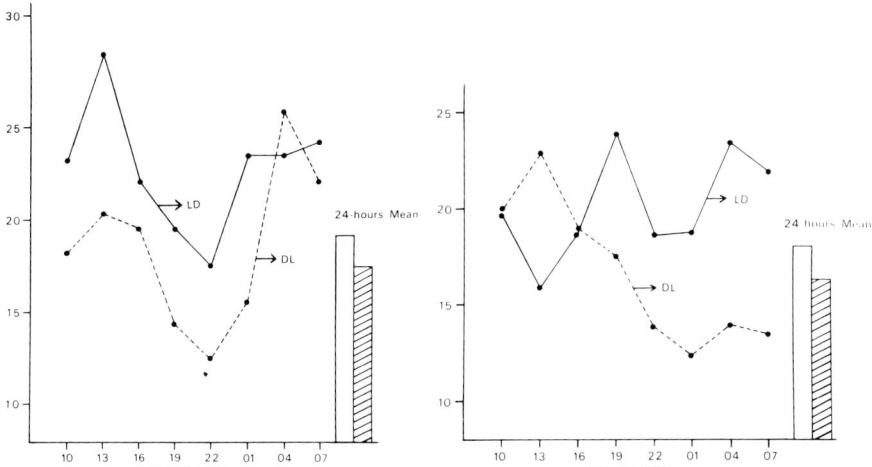

Figure 4. Semiquantitative evaluation of acid phosphatase.

Figure 5. Semiquantitative evaluation of B-glucuronidase.

B-glucuronidase. In both LD and DL conditions there is a clear circadian rhythm for B-glucuronidase. In LD rats the enzyme activity oscillated at a higher amplitude than in those maintained in DL, and the difference between the two was statistically significant.

In LD rats the peak activity was reached at 1900 hours, and the trough occurred at 1300; in DL the peak was at 1300 and the trough at 0100. Therefore, in DL the peak and trough occurred exactly 12 hours apart, whereas in LD this difference was only 6 hours. When the enzyme activity is plotted, the curve is biphasic for the animals in LD but monophasic for those in DL. B-glucuronidase oscillated above its 24-hour mean from 1000 to 1900 hours, i.e., during the light phase; and it was below the 24-hour mean during the dark phase (Fig. 5).

The peaks and troughs of acid phosphatase and B-glucuronidase are compared in Figure 6.

A. In LD schedule:
  1. The peak of acid phosphatase activity conicides with the trough of B-glucuronidase at 1300 hours.

2. The trough of acid phosphatase activity (2200 hours) occurred 3 hours after the peak of B-glucuronidase (1900 hours). Thus both the enzymes are present in the lysosomes, but their biodisposability depends on circadian phase. When the biodisposability of acid phosphatase is maximum, that of B-glucuronidase is minimum. This in turn will influence the function of the cells.

B. In DL schedule:
1. The peak of acid phosphatase activity occurred at 0400 hours and remained above the 24-hour mean level until 0700 hours. The peak of B-glucuronidase occurred at 1300.
2. The trough for acid phosphatase was at 2200 hours and for B-glucuronidase at 0100 hours, i.e., 3 hours later.

Thus in LD, with respect to the biodisposability of the substrate, the two enzymes are almost synchronized. The reversed lighting schedule acted as a better synchronizer than LD, at least as far as the biodisposability of these enzymes is concerned.

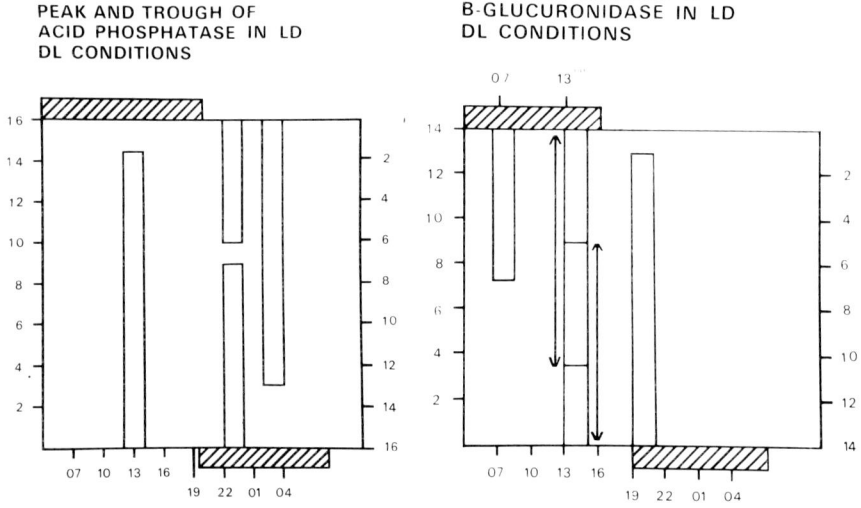

Figure 6.

Esterase at pH 5.0. At pH 5.0 both the lysosomal and microsomal enzymes are stained, and therefore the color intensity is comparatively greater than at pH 7.4 or 8.0 (Fig. 7).

In LD, esterase at pH 5.0 has two distinct peaks, one at 1000 hours and the other at 0100. The trough occurred between 1600 and 2200. The rhythm is completely desynchronized in DL with a sharp peak at 1000 hours which conicides with the one seen in LD; and there is a trough at 0700 hours. There are a few small peaks in between, and this suggests a damping of the rhythm. When the 24-hour mean values were compared, a reduction in total enzyme activity was noted in DL, but this difference is not statistically significant (Fig. 7).

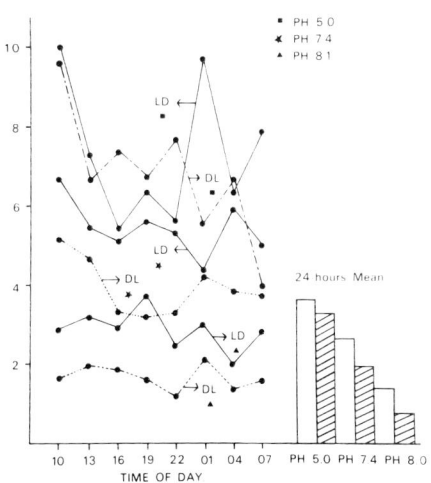

Figure 7. Semiquantitative evaluation of esterase.

Esterase at pH 7.4. At this pH, only the lysosomal esterases are reactive; thus the intensity of reaction is less than at pH 5.0.

In LD peaks can be seen at 1000, 1900 and 0400 hours (Fig. 7). In LD the enzyme activity is more or less synchronized. There is a sharp peak visible at 1000 hours and a flat trough between 1600 and 2200. Another small rise occurs at 0100. In DL, lysosomal esterase activity is similar to that seen at pH 5.0 in DL conditions (Fig. 7).

Esterase at pH 8.0. According to Holt (1958), only organophosphate-resistant esterase is stained at pH 8.0. Therefore the intensity of the reaction is very, very low. This can be seen in the histogram (Fig. 7). The esterase activity is only about 20% that at pH 5.0 and 50% that at pH 7.4. Because of the very low intensity of color developed, the distribution is almost equal at all time points. In DL conditions, there was even less activity.

Maximum enzyme activity occurred in LD at 1900 hours and in DL at 0100 hours. The enzyme reaction, in general, oscillated around the mean level; thus a very distinct circadian rhythm was not seen.

By changing the light schedule from LD to DL, the esterase activity at pH 5.0 was not affected from a chronobiological point of view. The activity oscillated at a lower range than in LD (Fig. 7). Either six-days time is not sufficient for this particular enzyme to phase shift and synchronize, or its quantity is so low that its biodisposability remains almost the same throughout the day.

DISCUSSION

Histoenzymological studies on three lysosomal enzymes of the liver show that all followed a circadian rhythm. Although all are present in the lysosomes, the biodisposability of their substrates is time dependent, each with its own time for peak activity. In this respect they manifest a heterogeneity. The heterogeneity of liver lysosomes on the basis of their chemical nature has already been reported by several workers (Novikoff, 1961; De Duve, 1963; Rahman et al, 1967). Bhattacharya and Mayersbach (1976) reported heterogeneity on the basis of size, distribution pattern and number of lysosomal granules. The present study further shows that these enzymes are not only heterogeneous with respect to their time-dependent biodisposability but also in their ability to synchronize with the light-dark schedule. Acid phosphatase, the marker enzyme for the lysosomal study, showed a shift of its peak 9 hours to the left; although the trough remained at the same time (Fig. 6). The peak for B-glucuronidase shifted by 6 hours to the right in the DL schedule and coincided with the LD peak of acid phosphatase. The trough for B-glucuronidase also shifted by 9 hours to the right (Fig. 6). The $180°$ shift of the LD cycle had a stronger

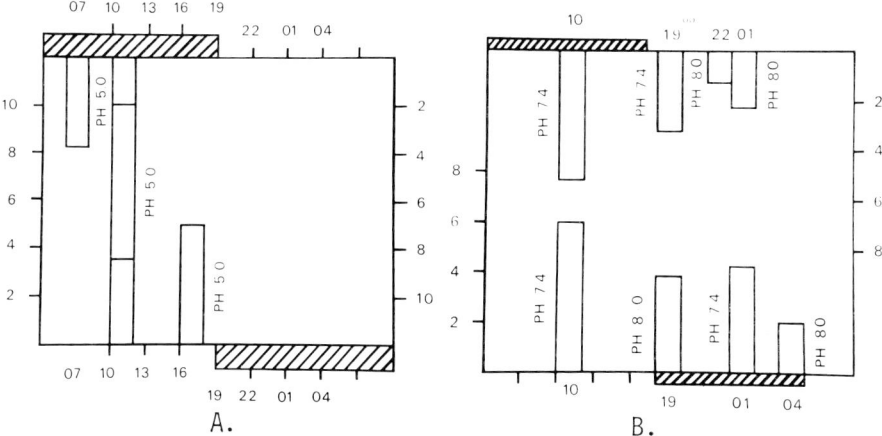

Figure 8. Peaks and troughs of esterase in LD and DL conditions. A, esterase at pH 5.0. B, esterase at pH 7.4 and pH 8.0.

effect on esterase activity. It did not disturb the rhythm of esterase at pH 5.0. The peak remained at 1000, but the trough shifted (Fig. 8a). At pH 7, the DL cycle could impose a circadian rhythm on the otherwise ultradian frequency of the enzyme. The peak activity occurred at 1000 (Fig. 8b). The DL schedule did not have much of an effect on the rhythmicity of reactive esterase at pH 8.0, excepting that the total activity was lowered.

The three lysosomal enzymes responded to the shift from LD to DL quite differently. The acid phosphatase reacted the most with a 9-hour shift in its peak activity, and B-glucuronidase shifted 6 hours. The DL schedule disturbed the rhythm of some esterases which were not able to resynchronize even after six days of acclimatization. Thus with respect to their different responses to the new light conditions, the lysosomes once more showed their heterogeneity. Some rapidly phase shifted to this new living condition, whereas others were still in the process of adjustment.

Many authors have already reported that some biological variables have the property to adjust swiftly to new environmental conditions after a short acclimatization period, whereas others need more time (Mayersbach et al, 1975; Scheving et al, 1974; Halberg and Nelson, 1966). The

present report has shown that six different enzymes that were demonstrated histochemically in liver lysosomes reacted differently to a new lighting regimen. That is, the time for maximum biodisposability of these enzymes shifted with the 180° change in the light-dark schedule, but the shift was only partial. In conclusion, it can be said that:
1. each enzyme studied has its own peak for biodisposability,
2. this peak is time dependent,
3. lysosomal enzymes are heterogeneous in nature with respect to their size, number and distribution pattern, and
4. lysosomal enzymes are heterogeneous in nature with respect to their ability to synchronize to a light-dark cycle.

Table 2. Semiquantitative evaluation of acid phosphatase and B-glucuronidase; Vitatron scanning.

Time	Acid phosphatase (cms)		B-glucuronidase (cms)	
	LD	DL	LD	DL
1000	11.15	9.17	9.93	10.0
1300	14.55	10.20	7.96	11.5
1600	11.10	9.80	9.40	9.5
1900	9.80	7.20	12.03	8.8
2200	8.80	6.25	9.30	7.0
0100	11.80	7.80	9.40	6.25
0400	11.80	13.0	11.25	7.0
0700	12.18	11.10	11.01	6.8
Mean	11.39	9.3	10.03	8.35
S.D.	± 1.70	± 2.20	± 1.31	± 1.87

Table 3. Semiquantitative evaluation of esterases from rat liver lysosomes; Vitatron scanning.

Time	Esterase (cms)					
	pH 5.0		ph 7.4		pH 8.0	
	LD	DL	LD	DL	LD	DL
1000	9.94	9.63	6.15	5.18	2.85	1.63
1300	7.3	6.66	5.45	4.65	3.18	1.96
1600	5.4	7.36	5.11	3.33	2.9	1.87
1900	6.38	6.76	5.63	3.2	3.71	1.63
2200	5.55	7.70	5.03	3.30	2.45	1.22
0100	9.2	5.56	4.3	4.22	3.01	2.10
0400	6.43	6.65	5.35	3.88	2.0	1.35
0700	7.85	3.80	5.01	3.76	2.81	1.55
Mean	7.25	6.76	5.25	3.94	2.86	1.66
SD	±1.65	±1.6	±0.53	±0.70	±0.50	±0.30

REFERENCES

Barka T, Anderson PT (1963). "Histochemistry Theory, Practice and Bibliography." New York: Harper & Row, p 243.

Bhattacharya R (1969). Chronobiological changes in rat liver enzymes. IInd Int Conf Expt Clin Chronobiology, Florence, Italy.

Bhattacharya R (1976). Application of histochemical technique in chronobiology. A study on rat liver lysosomes. 5th Int Conf Histo Cytochem Romania.

Bhattacharya R (1977a). Circadian difference of lysosomes and their enzymes. Nova Acta Leopoldina 46:171.

Bhattacharya R (1977b). Circannual changes of liver lysosomes of rat living in constant environmental conditions. Chronobiologia 4:131.

Bhattacharya R (1978). Temporal oscillation of cardiac muscle enzymes in rat. J Mol Cell Cardiology Suppl 1:9.

Bhattacharya RD (1980). Circadian rhythm of B-glucuronidase. Experientia 36:74.

Bhattacharya R, Mayersbach H von (1976). Histochemistry of circadian changes of lysosomal enzymes in rat liver. Acta Histochim 16:109.

Bhattacharya R, Mayersbach H von (1975). Effect of starvation on circadian rhythm of hepatic lysosomal enzymes in rat liver. Chronobiologia Suppl 1:8.

Bhattacharya RD, Mayersbach H von (1979). Circadian and seasonal rhythm of B-glucuronidase in rat liver. Chronobiologia 6:79.

Bhattacharya R, Wegmann R (1971). Effect of circadian rhythm on skeletal muscle enzymes. 7th Int Cong Anatomy, Leningrad.

Bhattacharya R, Wegmann R (1972). Difficulte dans les etudes de morphologie dans le recherche sur le rhythm circadien. Journe des etudes de methodologie dans le recherche circadien. Louvain, Belgium.

Bhattacharya RD, Wegmann R (1980). Liver respiratory enzymes and circadian rhythm. A histochemical study. Int J Cell Mol Biol (in press).

De Duve C (1963). General properties of lysosomes. In Reuck AVSD, Camerson (eds): "Ciba Foundation Symposium on Lysosomes," p 1.

Halberg F, Nelson W (1966). Phase relations of circadian rhythm: animals. In Altman PL, Dittmer DS (eds): "Environmental Biology," Bethesda, MD: Fed Soc Exp Biol Med, p 586.

Hayashi M (1964). Distribution of B-glucuronidase activity in rat tissues using the naphthol AS glucuronidase hexazonium pararosanilin method. J Histochem Cytochem 12:659.

Holt SJ (1958). Studies in enzyme cytochemistry V. An appraisal of indigogenic reactions for esterase localization. Proc Roy Soc 148:520.

Mayersbach H von (1967). Seasonal influences on biological rhythm of standardized laboratory animals. In Mayersbach H von (ed): "The Cellular Aspects of Biorhythms," Berlin: Springer, p 87.

Mayersbach H von, Philippens KMH, Scheving LE (1975). Light - a synchronizer of circadian rhythms. Proc XXI Int Conf Chronobiology, p 503.

Novikoff AB (1961). Lysosomes and related particles. In Brachet J., Mirski AE (eds): "The Cell," p 424.

Rahmann YE, Howe JF, Nance SL, Thompson JF (1967). Studies on rat liver ribonuclease. 2. Zonal centrifugation of acid ribonuclease implications for the heterogeneity of lysosomes. Biochem Biophys Acta 146:484.

Scheving LE, Mayersbach H von, Pauly JE (1974). An overview of chronopharmacology. J Europ Toxicol 7:203.

Uchiyama Y, Mayersbach H von (1979). Study on the rhythmic changes of lysosomal enzyme activities in rat liver parenchymal cells using ultracytochemistry. Chronobiologia 6:166.

This work was carried out at the Institute of Anatomy I, Hannover Medical School, West Germany; H. von Mayersbach, Director.

RHYTHMS IN SUSCEPTIBILITY

CIRCADIAN SYSTEM AND TERATOGENICITY OF CYTOSTATIC DRUGS

Ingrid Sauerbier

Medizinische Hochschule Hannover
Department Anatomie
Karl-Wiechert-Allee 9, D 3000 Hannover 61, FRG

Drug-induced teratogenesis in mammals has been the goal of many studies during the past few years. Many chemical compounds are known to cause certain types of malformations or hazards in the embryo. According to these studies it is generally accepted that the teratogenic effect depends upon the nature of the drug, the dosage and the date of gestation at which a drug is administered (cf. Nishimura and Tanimura, 1976).

Until now, little has been done with respect to the evaluation of the teratogenicity of drugs relative to the time of day of their administration on a certain gestational day (Isaacson, 1959; Clayton et al., 1975; Schmidt, 1977). The present study was designed to examine the teratogenic effects of two alkylating agents, being nitrogen mustards (cyclophosphamide and Th-R), with regard to the time of day at which the drug is given on day-12 of gestation, and to compare the circadian (= about 24-h rhythm) response rhythm in drug susceptibility. Both aspects could be of great importance when screening teratogenic potentialities of a new drug.

MATERIALS AND METHODS

NMRI/HAN mice (SPF-grade, averaging about 28g at the time of mating) were kept under highly standardized conditions (temperature 22±1°C, relative humidity 55%, artificial light-dark regimen, light from 06.00 to 18.00), standard food (Altromin, Lage,FRG) and water ad lib. After an acclimatization period of at least 2-3 weeks, groups of 4 females were caged with one male according to three different limited timed matings: 07.00 - 09.00; 19.00 - 21.00 or 17.00 - 07.00. Mating was confirmed by the presence of a vaginal plug. On the 12th day of gestation, groups of 10 animals received a single intraperitoneal injection of cyclophosphamide (Endoxan^R, Asta Werke, Bielefeld, FRG; dissolved in distilled water) or Th-R (an experimental N-mustard compound, Thiemann, Lünen, FRG; suspended in 1% methylcellulose) at one of four different day-times (07.00, 13.00, 19.00, 01.00). Control animals were injected with an equivalent volume of solution medium (1o ml/kg) or remained untreated. The exposure time points were calculated from the time of conception.

The pregnant mice were sacrificed on the 18th day of gestation and the implantation sites in each uterine horn were counted. Live fetuses were weighed individually; after examination for external malformations, fetuses were prepared for staining the skeleton with alizarin red S (Dawson,1926). For statistical evaluation the Student's t-test and the F-test were used.

RESULTS

The effect of a single therapeutic dose of cyclophosphamide (20 mg/kg, LD 10) was quantitatively and qualitatively related to the circadian phase of drug administration. Maximum teratogenicity was produced by maternal treatment at 07.00. There was a 89.9% incidence of malformations compared with 50.9% due to maternal treatment at 01.00. The magnitude of the teratogenic effects induced by cyclophosphamide varied in between, being generally more severe when administration was performed during the light (i.e., **inactive** phase). Gross examination of fetuses revealed open eyes, aphakia, cleft palate, exencephaly and kinky tail. Skeletal anomalies most common included ossification defects of the vertebral column, sternum, ribs and limb**s (Fig. 1).**
It is interesting to note that within a given litter of each treatment group all fetuses were similarly affected, corresponding to the teratogenic response at the circadian phase (time) of injection.

The teratogenic effect of Th-R appeared to be also modified by the time of day that the drug was administered (Fig. 1 and 2). Maternal treatment with a single dose of

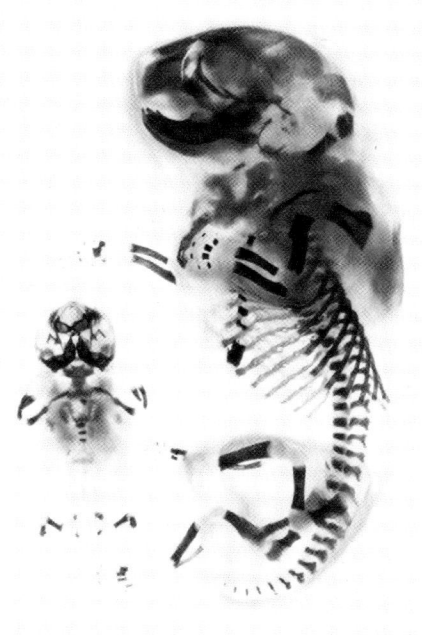

Fig. 1. Teratogenic effects in mouse fetuses due to maternal treatment with 2 mg/kg Th-R at 07.00 (right) and at 19.00 (left).

2 mg Th-R per kg body weight (LD 10) at the onset of the dark period (19.00) resulted in a strong growth retardation, and 87% of the fetuses exhibited malformations (e.g., micrognathia, absence or nonossification of ribs, sternal malformations, hypoplasia and aplasia of vertebrae, limb and skull abnormalities). In contrast, when the drug was given at the beginning of the light period (07.00), no growth inhibition was apparent, and in only 6.2% of the fetuses slight malformations were observed.

Fetal weight was also markedly affected by treatment of either cyclophosphamide or Th-R. Thus an increase in weight correlated with a decrease in severity of malformation, e.g. Th-R: 0.46 g at 19.00 vs. 1.2 g at 07.00.

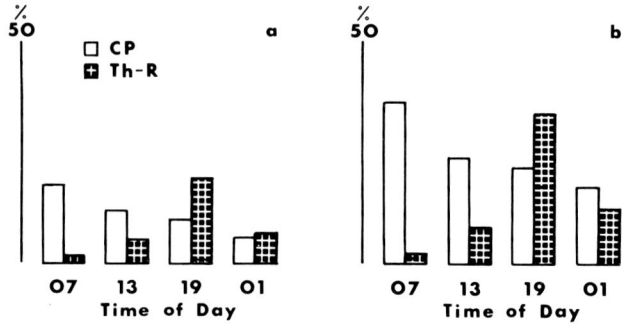

Fig. 2. Number of cyclophosphamide or Th-R induced malformations per fetus, (a) one malformation per fetus and (b) more than one.

Control fetuses **had no skeletal anomalies or had a markedly low incidence of malformations** ranging in the rate of spontaneous malformations as cited in literature (Frohberg, 1978).

DISCUSSION

For several drugs and poisons circadian phases of susceptibility have been identified (Scheving et al., 1974; v. Mayersbach, 1976). Moreover, such varying toxic effects on male and non-pregnant female mice, as influenced by the time of day of drug administration , have been shown for cyclophosphamide (Haus et al., 1974; Cardoso et al., 1978; Rose et al., 1978) and for Th-R (Anagnou et al., 1979). There is now additional evidence that confirm the circadian dependence of cyclophosphamide and Th-R induced teratogenesis.

The teratogenic action of cyclophosphamide in mice has been noted previously, **e.g., by Gibson et al. (1968) and Gebhardt (1970).** Moreover, day 11-12 of pregnancy was found to

be the most effective day to induce malformations. In
addition, in the present study it could be demonstrated that
the teratogenic response to cyclophosphamide follows a
definite circadian pattern. The same holds true for Th-R.
However, both drugs differ with respect to their mean
embryotoxicity as well as in their phasing (time) and
amplitudes of maximum and minimum sensitivity.

At the beginning of the light phase, cyclophosphamide
exhibits strongest teratogenic effects when **the least effect**
of Th-R occurs. Highest teratogenicity of Th-R has been
observed at the onset of darkness. It should be noted that
this maximal teratogenic effect of Th-R is much more
pronounced, compared to the maximal effect of cyclophosphamide which in turn causes in the mean more malformations.

With regard to the circadian pattern of cyclophosphamide, the present results are in general agreement with those
described by Schmidt (1977). However, they differ with
respect to the mean embryotoxicity and the lack of specifi-
cation of malformations. In part, the discrepancy is due to
the use of different stocks of mice in the two laboratories
and to differences in the experimental design, e.g. light
conditions.

The amplitude differences in the teratogenicity of
both drugs are paralleled by their circadian lethal effects
on male and non-pregnant female mice (Anagnou et al., 1979).
Here, too, a higher amplitude rhythm has been found for
Th-R. Further, the time points of highest and lowest toxicity
of cyclophosphamide and Th-R coincide well with the occurrence
of their maximum and minimum teratogenicity. This parallelism
between toxicity and teratogenicity suggests that t e
teratogenic effect may be directly caused by the two agents,
without ruling out the possible importance of the amount
of drug reaching the fetus.

Moreover, frequency, severity and combination of malformations induced by both drugs, seasonally vary when **the
circadian studies were repeated** under identical conditions
during different months of two consecutive years. Seasonal
modifications of the mean embryotoxicity, being associated
with a principally unchanged circadian pattern, are highest
during spring and summer and lowest during winter, although
the animals were constantly kept under standardized artificial environmental conditions. Such seasonal variations in

susceptibility of non-pregnant animals to drugs have already been described (v. Mayersbach, 1976; Philippens, 1976). In non-rhythmic studies a few facts are also reported with respect to seasonal variation in frequency of the drug induced malformations. Thus, frequency of cortisone-induced cleft palate in mice as well as the entire array of malformations produced by maternal vitamin A administration are found to be much greater in the winter months than in the summer months (Kalter, 1959; Kalter et al., 1961).

In summary, the time of day which the drug is given is expected to influence the teratogenic outcome insofar as the embryo has different sensitivities on the same day. Therefore rhythms should be taken into account when evaluating the effects of a teratogen. Further studies are needed to separate out the influence of maternal (?) circadian rhythms from varying and ongoing developmental sensitivity of the embryo.

SUMMARY

Pregnant mice were injected once with cyclophosphamide (20 mg/kg) or Th-R (N-mustard, 2 mg/kg) at one of four different day-times (07.00, 13.00, 19.00, 01.00) on the 12th day of gestation. In a therapeutic dose both alkylating neoplastic agents induced teratogenic effects relative to the maternal circadian cycle. The highest incidence of malformations due to maternal treatment with cyclophosphamide was found to be associated with the dark to light transition (07.00), whereas the lowest occurred at 01.00. In contrast, the teratogenic action of Th-R was strongest at the onset of darkness (19.00) and lowest at 07.00. The mean embryotoxicity of both compounds was subjected to seasonal modifications, being highest during spring and summer and lowest during winter. - The observations indicate that for the evaluation of the teratogenic potentialities of drugs, rhythms cannot be neglected.

REFERENCES

Anagnou J, Mayer D, v. Mayersbach H (1979). Circadian toxicity of cytostatic drugs. Chronobiologia (in press).
Cardoso S S, Avery T, Venditti J M, Goldini A (1978). Circadian dependence of host and tumor responses to

cyclophosphamide in mice. Eur J Cancer 14: 949.
Clayton D L, Mc Mullen A W, Barnett C C (1975). Circadian modification of drug-induced teratogenesis in rat fetuses. Chronobiologia II (3): 210.
Dawson A B (1926). A note on the staining of the skeleton of cleared specimens with alizarin red A. Stain Technol. 1: 123.
Frohberg H (1978). Zur Übertragbarkeit toxikologischer Versuche auf den Menschen. In Weihe W H (ed): "Das Tier im Experiment", Bern: Huber, p 175.
Gebhardt D O E (1970). The embryolethal and teratogenic effects of cyclophosphamide on mouse embryos. Teratology 3: 273.
Gibson J E, Becker B A (1968). The teratogenicity of cyclophosphamide in mice. Cancer Res 28: 475.
Haus E, Fernandes G, Kuhl J F W, Yunis E J, Lee J D, Halberg F (1974). Murine circadian susceptibility rhythm to cyclophosphamide. Chronobiologia I: 270.
Isaacson R J (1959). An investigation of some of the factors involved in the closure of the secondary palate. Thesis Ph D University of Minnesota, Minneapolis, Minnesota.
Kalter H (1959). Seasonal variation in frequency of cortisone-induced cleft palate in mice. Genetics, 44: 518.
Kalter H, Warkany J (1961). Experimental production of congenital malformations in strains of inbred mice by maternal treatment with hypervitaminosis A. Amer Path J, 38: 1.
Mayersbach H v. (1976). Time: a key in experimental and practical medicine. Arch Toxicol 36: 185.
Nishimura H, Tanimura T (1976). Clinical aspects of the teratogenicity of drugs. Excerpta Medica, Amsterdam.
Philippens K M H (1976). The manipulation of circadian rhythms. Arch Toxicol 36: 277.
Scheving L E, Mayersbach H v., Pauly J E (1974). An overview of chronopharmacology. J Eur Toxicol 7: 203.
Schmidt R (1977). Zur zirkadianen Modifikation der teratogenen Wirkung von Cyclophosphamid. 1. Mitteilung. Biol Rundschau 15: 314.

CIRCADIAN STAGE DEPENDENCE IN RADIATION: RESPONSE OF DIVIDING CELLS IN VIVO

NORMA H. RUBIN

Dept. of Anatomy, Univ. of Arkansas for Medical Sciences
Little Rock, Arkansas 72201
Present Address: University of Texas Medical Br.
Div. of Cell Biology, Galveston, Texas 77550

INTRODUCTION

Ionizing Radiation and the Circadian Susceptibility-Resistance Cycle

The marked morphological changes in tissues produced by ionizing radiation were first described in the early 1900's. The arrest or delay of mitosis in replicating cells due to X-rays was clearly recognized then, and in 1906 Bergonie and Tribondeau formulated their law which is central to the radiotherapy of cancer: "The sensitivity of cells to irradiation is directly proportional to their reproductive activity and inversely proportional to their degree of differentiation." Post-irradiation changes within the cell cover a wide spectrum, and it is unlikely that any one mechanism will be found to explain the varied cellular effects. Many investigators, however, study the effects of radiation in both in vitro and in vivo systems. Most have been unaware of the importance of the circadian system in an animal's response to ionizing radiation.

In the 1960's it was clearly demonstrated by several investigators that a toxic dose of radiation given to one group of animals at one time would kill more than the same dose given at another time. Halberg et al later showed that 500 Roentgens killed only 30% of the animals when given near the end of the light period, but killed 100% when given at the beginning of the light period. This circadian response could be partially shifted in animals which had been standardized to an inverted light-dark (LD) cycle (for review see Haus et al, 1974).

Chronotolerance to radiation has been reported not only at the level of survival or death of the animal but also at the cellular level. A circadian variation in radiation response has been demonstrated in liver regeneration, blood leukocytes, intestinal crypt cell survival, and spleen colony forming units (Ueno, 1968; Vacek and Rotkovska, 1970; Pizzarello and Witcofski, 1971; Lappenbusch, 1972; Barbason, 1974; Hendry, 1975; Becciolini et al, 1979). These findings document a concept fundamental to chronobiology, that of hours of changing susceptibility and resistance of an organism to perturbations such as drugs, physical agents, and toxins. There are numerous examples of these cycles (Halberg et al, 1958; Reinberg et al, 1965; Scheving et al, 1976; Sturtevant et al, 1976). In fact, almost any agent can be shown to have different effects at different stages in the circadian cycle. Since the organism is not actually in homeostasis in the classical sense at all, but rather is in a continuous state of physiological flux, it therefore will react differently to an identical stimulus at different stages of its circadian system. All drugs or agents which have the potential to affect dividing cell populations, including ionizing radiation, would be expected to demonstrate the concept of hours of changing susceptibility and resistance simply because cell division itself also is rhythmic.

Circadian Rhythm in Cell Division

The circadian rhythm in cell division in vivo, first described in 1917 by Fortuyn-van Leyden, has been abundantly documented (for review see Scheving, 1959). Today a circadian rhythm in cell division can be demonstrated in any mitotically active normal tissue from properly standardized animals, including man. An example is seen in the rodent corneal epithelium, which has been extensively investigated over the past twenty years as a model for the circadian variation in cell division (Vasama and Vasama, 1958; Alov, 1959; Kosichenko, 1960; Vinogradov, 1961; Scheving and Pauly, 1967 a,b; Burns and Scheving, 1975; Cardoso et al, 1978). Its rhythm in cell division exhibits one of the larger changes in amplitude over a 24-hour (h) period (Fig. 1). Often the peak in mitotic index is more than twenty times the trough value. The wide variation around a 24-h mean is an admonition of the hazard of sampling at only one or two random times in the 24-h cycle, as is the practice in many laboratories.

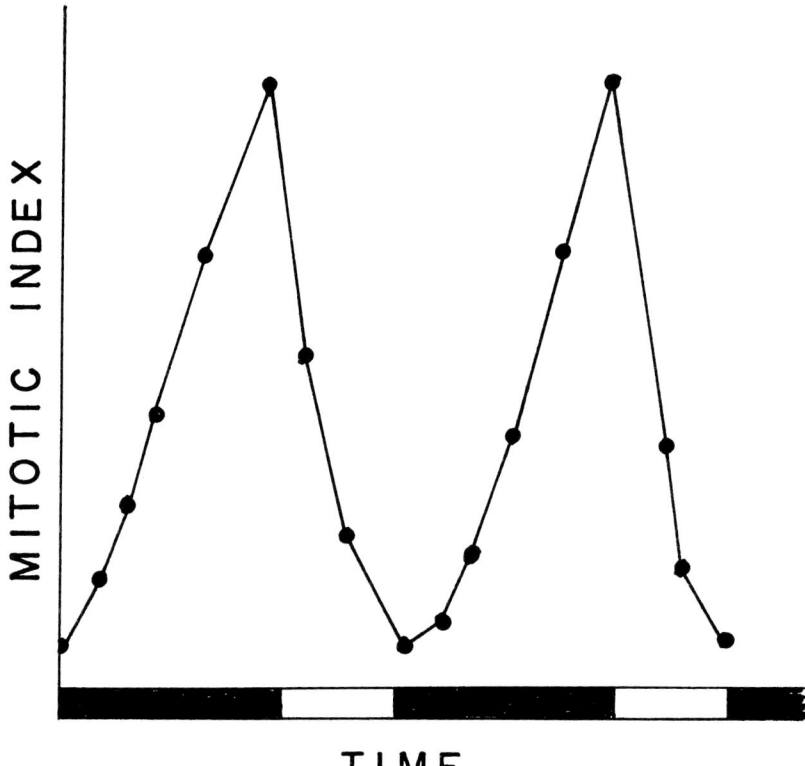

Figure 1 Circadian rhythm in cell division in mouse corneal epithelium. Peak and trough in mitotic index (ordinate) are synchronized to the LD schedule (abscissa) and therefore occur at predictable intervals of time. Darkness in indicated by dark horizontal bars.

The maximum and minimum values do not occur at random but are predictable from one experiment to another under constant stndardization conditions. The corneal epithelium is only three to twelve cell layers thick (Elgjo, 1969 ; Rehbinder, 1978); and the stroma is relatively thin, allowing one to analyze the mitotic index from whole mount preparations. The determination of the mitotic index is easily performed by examining a predetermined number of microscopic fields.

The corneal epithelium is therefore a practical and advantageous tissue for studying some of the effects of ionizing radiation on replicating cells. One specifically defined effect is radiation-induced mitotic delay.

Radiation-Induced Mitotic Delay

Mitotic delay is a fundamental concept of radiation biology described in many textbooks as a temporary G_2 block in cells approaching mitosis. The next division is delayed for a time reported to be dose and cell cycle dependent. An integral part of the concept is the mitotic rebound which occurs when the arrested cells recover from the G_2 block, reenter the pool of dividing cells, and add their numbers to those normally entering mitosis at that time. This concept was derived from studies on asynchronously dividing cells growing in culture, which are characterized by a constant level of mitotic index. Available evidence suggests that there are no asynchronously dividing cells in vivo. It is therefore questionable whether it is valid to attempt to extrapolate data derived from in vitro studies to in vivo systems without considering the circadian fluctuation in cell division.

Several investigators previously have reported a relationship between the circadian rhythm in mitosis in the epithelium of the cheek pouch and the time when tritiated thymidine was injected into hamsters (Gibbs and Casarett,1969,1971; Møller et al,1974). Barbason (1974, 1976) examined in vivo mitotic delay subsequent to radiation in the regenerating rat liver. A more extensive attempt to relate the effect of circadian rhythmicity and radiation response in dividing cells was reported by Lesher and Lesher (1970). They administered various single whole-body doses to mice, although at only one circadian stage, and monitored the number of cells in mitosis and the uptake of tritiated thymidine in the crypt cells of the duodenal epithelium. However, their results appear to support the classical concept of mitotic delay with rebound.

Using the murine corneal epithelium, I have investigated the influence of the circadian rhythm on the radiation response of the dividing cells in this tissue. More comprehensive data from these experiments have been submitted for publication elsewhere.

METHODS AND MATERIALS

In all experiments male CD_2F_1 or $B_6D_2F_1$ mice, approximately four to six weeks old, were standardized for three weeks to a LD schedule of 8 h light and 16 h darkness. Food and water were freely available and the temperature was maintained at 22 ± 2^oC. Mice were irradiated in a cesium-137 small animal irradiator; an equal number of mice were sham-irradiated. A dose-response experiment and an isodose experiment were designed to explore for a circadian influence on the effect of ionizing radiation on the mitotic index in the corneal epithelium.

In the dose-response study one of three doses (200, 600, or 1000 rads) was given to each of three groups of mice at the same circadian stage, the time of transition from dark to light. Irradiated and sham-irradiated mice were killed every 3 h for 48 h after irradiation, and the eyeballs were immediately removed. The mitotic index was determined on the corneal epithelium as described by Scheving and Pauly (1967).

The isodose experiment was designed so that identical doses (600 rads) of radiation were given to six groups of mice, each at a different circadian stage. Six other groups served at sham-irradiated controls. The times of irradiation were 0 h, 6 h, 9 h, 12 h, 18 h and 21 h after "lights on." Four of these groups were treated during the dark hours of their LD schedule; they were exposed only to dim red light throughout the irradiation or sham-irradiation procedure. Samples of irradiated and sham-irradiated animals were killed every 3 h for 48 h after irradiation and the corneal epithelium analyzed for mitotic index as described before.

RESULTS AND DISCUSSION

Isodose Experiments

There were six groups of sham-irradiated mice in the isodose study. Each was sham-irradiated at a different circadian stage. The circadian rhythm of the mitotic index was disturbed little if any by the procedure, as shown diagramatically in figure 2a. Each demonstrated the characteristic high-amplitude fluctuations reported by others with a variation of more than twenty-fold over each 24-h period.

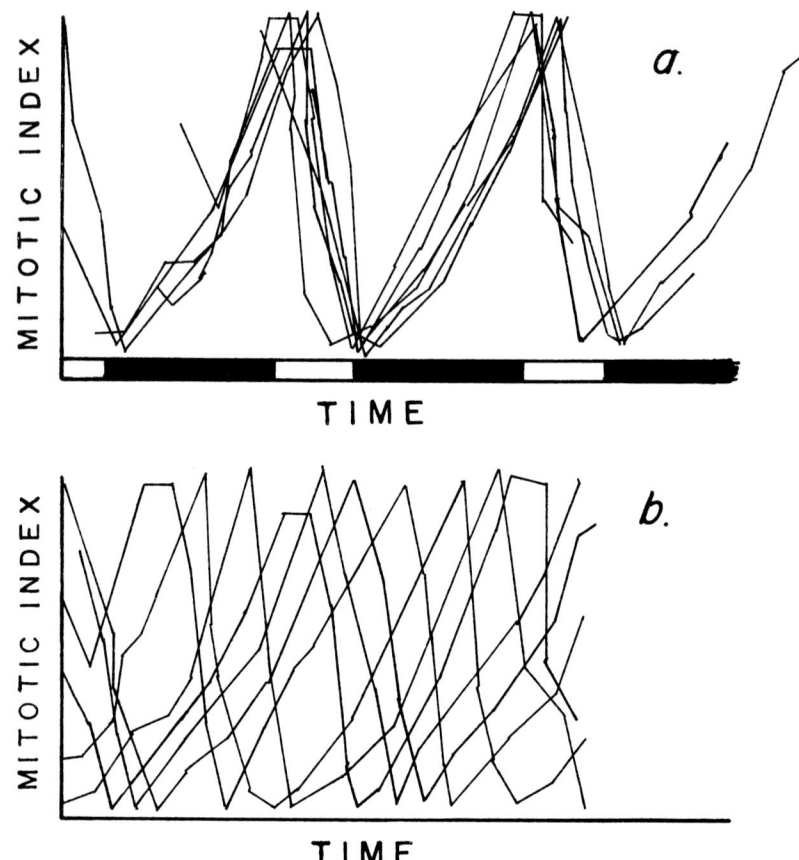

Figure 2   Mitotic index rhythms (control animals) from six isodose experiments plotted with reference to the LD schedule (abscissa) (a) or with reference to the number of hours after sham-irradiation (abscissa) (b). There is no apparent difference in the six control rhythms in 2 a; however 2b shows an apparently chaotic picture.

With the sampling intervals used, the trough, where mitosis is almost negligible, occurred in every case 9 h after lights went on, or about 1 h into the dark span. In every case the peak occurred about the time of transition from dark to light. These results show clearly how one-time-point sampling can be misleading. If one disregards "time of day" when each experiment was done (as is frequently done) and plots the identical control data by "hours after sham-irradiation," one can readily see that the data appear dissimilar (Fig. 2b). At the beginning of the experiment, the "control"value for the mitotic index could have been very low or very high. Of if one took a random control sample at 24 h after irradiation, values again could have been either low or high. Therefore the method of sampling at one time to avoid the effects of rhythms is not supported by these data.

There were six groups of irradiated mice in this study. Each was irradiated with the same dose (600 rads) but at a different circadian stage. However, the effect of 600 rads on the mitotic index in the corneal epithelium was markedly different in each case. Mitosis was inhibited in all experiments as early as 1½ h after irradiation, but the duration of mitotic inhibition did not appear fixed for one dose as one would have expected (Friedenwald and Sigelman, 1953; Machemer et al, 1968). The isodose experiments showed that mitotic delay induced by 600 rads varied depending upon the time it was given in the 24 h period. The length of delay ranged from 6 h to 15 h. In no case did the appearance of the recovery wave, which signalled release of the radiation-induced block, begin at a time other than during the daily increase in mitotic index in the control animals. No rebound was seen, a phenomenon consistently reported by others. Varying the time of the perturbation was not effective in forcing recovery during trough time in controls. Regardless of the time at which irradiation was administered, the entire wave of recovery occurred within the same time span as the controls. This recovery period was always promptly terminated at the same time the controls reached their lowest values for the mitotic index in the 24-h period.

In the further analysis of these data, four other notable differences were found in the response among the six experiments. The differences were variations in 1) span of the first post-irradiation recovery curve, 2) time span in number of hours after irradiation during which recovery transpired, 3) peak value in mitotic index of the first post-irradiation peak, and 4) area under the first post-irradiation curve, perhaps indicative of a variation in rate of mitosis. The data from the irradiated animals can be diagrammed two ways--either with regard to the "number of hours after irradiation," as is usually done in the literature (Fig. 3a), or with regard to the LD schedule(Fig. 3b). The

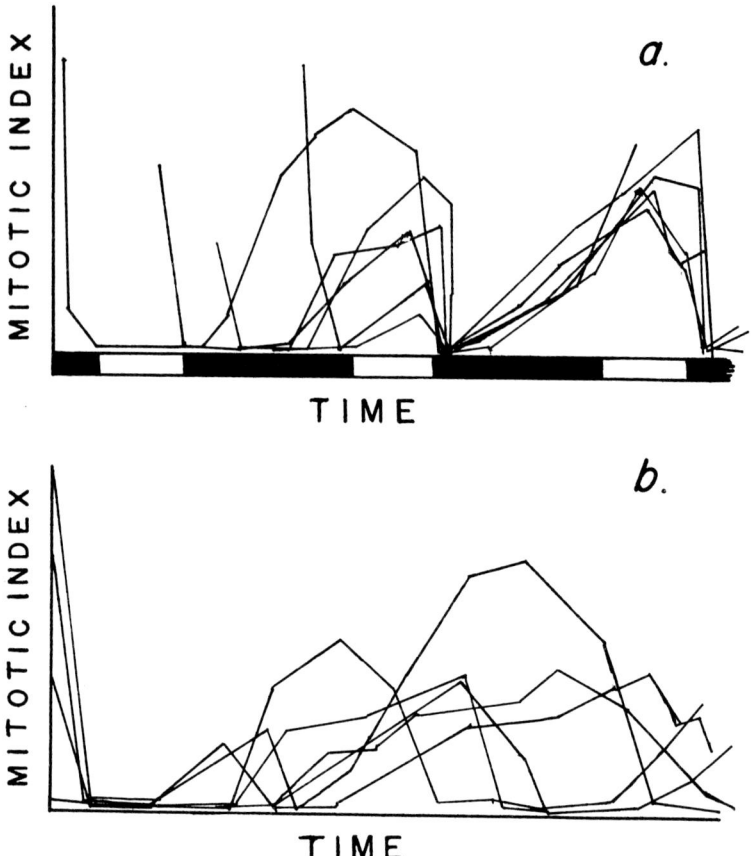

Figure 3 Mitotic index rhythms (irradiated animals) from six isodose experiments plotted with reference to the LD schedule (abscissa) (a) or with reference to the number of hours after irradiation (abscissa) (b). In (a) one can see a pattern to the radiation response but one is not evident in (b).

phasing of the curves appears quite similar when plotted by the LD schedule, whereas the interpretation of the data would be tenuous in the other case.

In fact, the epithelial cells of the mouse cornea demonstrate a circadian susceptibility-resistance cycle to radiation-induced mitotic delay and recovery. The data show that there is a generally increasing deleterious effect of the same dose on the first recovery wave when irradiation is performed at these times after lights on: 0 h, 6 h, 9 h, 12 h, 18 h, 21 h. The results suggest a substantial influence of the underlying circadian rhythm in cell division.

Dose-Response Experiment

The sham-irradiated animals again demonstrated the characteristic circadian rhythm of mitotic index in the corneal epithelium. In all iradiated animals the mitotic index dropped to essentially zero by the first sample time after irradiation. The length of time mitosis was inhibited appeared to be dose dependent. By 6 h after irradiation, the mitotic index in animals which has received the lowest dose (200 rads) already showed a significant increase. Meanwhile, the 600-rad group remained at trough level through 15 h, and the 1000-rad group through 18 h after irradiation.

Approximately 24 h after sham-irradiation, another peak in mitotic index occurred in the control animals. Again, a dose-response was seen at this time in the peak mitotic index attained in the irradiated animals; the 200-rad group reached the highest value, and the 1000-rad group appeared to be inhibited the most by the radiation. At the time of the next expected trough, 33 h after irradiation, the mitotic index in all three irradiated groups was again at trough level. The same pattern was evident in the beginning of the next post-irradiation curve beginning after the 33-h sample; that is, the group receiving the lowest dose appeared to "recover" to a greater extent than the group receiving the highest dose, with the middle-dose group showing intermediate recovery. Again, the results suggest a substantial influence of the underlying circadian rhythm in cell division.

The results of the data presented here further suggest that it is inappropriate to apply the results of *in vitro* studies on mitotic delay to an *in vivo* system. In one-celled organisms such as Tetrahymena, Euglena, and Gonyaulax, the initiation of DNA synthesis and mitosis have been shown to be synchronized to a circadian periodicity. However, this assumption is not unequivocally valid for isolated organs or cells of multicellular organisms. These cells came from an environment where they synchronized to signals from outside the cell that may have been borne by blood or lymph, signals which may no longer be present in cell cultures. So, mammalian cells in culture, isolated from normally present humeral influences, should not be expected to respond to experimental stimuli in a manner representative of the whole organism--whether one is considering radiation-induced alterations in cell kinetics or any other variable.

Reports in the literature consistently emphasize the correlation between cell-cycle position at the time of radiation and degree of cell killing or mitotic delay (Dewey and Humphrey, 1962; Puck and Steffan, 1963; Terasima and Tolmach, 1963; Sinclair, 1968; Dewey et al, 1971; Pizzarello and Witcofski, 1971; Highfield and Dewey, 1975). Radiation biologists and therapists have viewed the in vitro cell-cycle effect on mitotic delay as being of possible significance in radiation-therapy scheduling; however, their concepts have been derived without consideration of the circadian rhythm characterizing cell division in vivo. To study more precisely the phenomenon of mitotic delay in vivo, so that it possibly may be of clinical significance, both the circadian rhythmicity in cell division and the organism's overall circadian susceptibility-resistance cycle to radiation must be included as variables. Moreover, utilization of the cell cycle data thus derived, combined with use of synchronizing drugs and radiation, can lead to new approaches in the therapy of cancer in man.

SUMMARY

1) When the mouse corneal epithelium was irradiated with one dose of ionizing radiation, the effect on mitosis varied depending upon the time in the 24-h period when it was irradiated.

2) Release of the radiation-induced mitotic block, as measured by appearance of a recovery wave of mitotic cells, occurred only during the daily increase in mitotic index in the control animals.

3) The entire wave of recovery occurred within the same time as the controls, and no rebound was seen.

4) Mitotic delay was dose-dependent.

5) The results of these experiments emphasize the importance of considering the circadian system when studying cell division in vivo.

REFERENCES

Alov I A (1959). The mechanism of the diurnal periodicity of mitosis. Bull Exp Biol Med (USSR) 48: 1418.

Barbason H (1974) Influence of X-irradiation on the different stages of the cell cycle in regenerating rat liver. Virchows Arch (Cell Path) 16: 363.

Barbason H (1976) Influence of the circadian rhythm of cell division on the effect of X-irradiation in the regenerating rat liver. Int J Radiat Oncol Biol Phys 1: 911.

Becciolini A, Balzi M, Benucci A, Cremonini D, Franciolini F, Rizzi M (1979). Modifications in the small intestine after multiple irradiation starting at different hours of the day. Note 1: Modifications of cellular kinetic parameters. Chronobiologia 6 (2): 77.

Burns E R, Scheving L E (1975). Circadian influence on the wave form of the frequency of labeled mitoses in mouse corneal epithelium. Cell Tissue Kinet 8 (1): 61.

Cardoso S S, Philippens, K M H, Mayersbach H v (1978). The effect of cyclophosphamide upon mitoses in the cornea of rats, a circadian dependent effect. Eur J Cancer 14: 1037.

Dewey W C, Humphrey R M (1962). Relative radiosensitivity of different phases in the life cycle of L-P59 mouse fibroblasts and ascitic tumor cells. Radiat Res 16: 503.

Dewey W C, Miller H H, Leeper D B (1971). Chromosomal aberrations and mortality of X-irradiated mammalian cells: Emphasis on repair. Proc Natl Acad Sci 68(3): 667.

Elgjo K (1969). Cell renewal of the normal mouse cornea. Acta Pathol Microbiol Scand 76: 25.

Friedenwald J S, Sigelman S (1953). The influence of ionizing radiation on mitotic activity in the rat's corneal epithelium. Exp Cell Res 4: 1.

Gibbs S J, Casarett G W (1969). Influences of a circadian rhythm and mitotic delay from tritiated thymidine on cytokinetic studies in hamster cheek pouch epithelium. Radiat Res 40: 588.

Gibbs SJ, Casarett GW (1971). Cytokinetic effects of repeated X-irradiation in vivo in the presence of a circadian rhythm in mitotic activity. Radiat Res 48: 265.

Halberg F (1958). Physiologic 24-hour periodicity: General and procedural consideration with reference to the adrenal cycle. Z Vitam-Horm u Fermentforsch 10: 225 .

Haus E, Halberg F, Loken MK, Kim YS (1974). In Tobias CA, Todd P (eds): "Space Radiation Biology," New York: Academic Press, p. 435.

Hendry J H (1975). Diurnal variations in radiosensitivity of mouse intestine. Br J Radiol 48: 312 .

Highfield D P, Dewey W C (1975). Use of the mitotic selection procedure for cell cycle analysis: Emphasis on radiation-induced mitotic delay. Methods Cell Biol 9: 85 .

Kosichenko L P (1960). The character of the 24-hour periodicity of mitosis in the corneal epithelium of various laboratory animals. Bull Exp Biol Med (USSR) 49 (6): 98 .

Lappenbush, W L (1972). Effect of circadian rhythm on the radiation response of the Chinese hamster (Cricetulus griseus). Radiat Res 50: 600 .

Lesher J, Lesher S (1970). Effects of single-dose, whole-body ^{60}Co gamma irradiation on number of cells in DNA synthesis and mitosis in the mouse duodenal epithelium. Radiat. Res. 32: 510

Machemer R, Schuster R, Süchting P, Büttner H (1968). Das verhalten von DNS-synthese and mitosetatig-keit nach einwirkung von betastrahlen auf das corneal epithel. Strahlentherapie 136: 308 .

Møller U, Larsen J K, Faber M (1974). The influence of injected tritiated thymidine on the mitotic circadian rhythm in the epithelium of the hamster cheek pouch. Cell Tissue Kinet 7: 231 .

Pizzarello D J, Witcofski R L (1971). Regenerating rat liver: A good system for radiobiological studies. Radiology 100: 163 .

Puck T T, Steffan J (1963). Life cycle analysis of mammalian cells. I. A method for localizing metabolic events within the life cycle, and its application to the action of colcemide and sublethal doses of X-irradiation. Biophys J 3: 379 .

Rehbinder C (1978). Fine structure of the mouse cornea. Z. Versuchstierkd. 20: 28 .

Reinberg A, Ghata J, Sidi E (1965). Circadian reactivity rhythms of human skin to histamine or allergen and the adrenal cycle. J Allergy 36 (3): 273 .

Scheving L E (1959). Mitotic activity in the human epidermis. Anat Rec 135: 7 .

Scheving L E, Pauly J E (1967a). Effect of adrenalectomy, adrenal medullectomy and hypophysectomy on the daily mitotic rhythm in the corneal epithelium of the rat. In Mayersbach Hv (ed): "The Cellular Aspects of Biorhythms," Berlin: Springer-Verlag.167-174

Scheving LE, Haus E, Kuhl JFW, Pauly JE, Halberg F, Cardoso S (1976). Close reproduction by different laboratories of characteristics of circadian-rhythm in 1-beta-D-arabinofuranosylcytosine tolerance by mice. Cancer Res 36 (3): 1133.

Sinclair WK (1968). Cyclic X-ray responses in mammalian cells in vitro. Radiat Res 33: 620 .

Sturtevant FM, Sturtevant RP, Scheving LE, Pauly JE (1976). Chronopharmacokinetics of ethanol. II. Circadian rhythm in rate of blood level decline in a single subject. Naunyn Schmiedebergs Arch Pharmacol 293(3): 203 .

Terasima T, Tolmach L T (1953). Variations in several responses of Hela cells to X-irradiation during the division cycle. Biophys J 3: 11 .

Ueno Y (1968). Diurnal rhythmicity in the sensitivity of haemopoietic cells to whole-body irradiation of mice. Int J Radiat Biol 14(4): 307 .

Vacek A, Rotkovska D (1970). Circadian variations in the effect of X-irradiation on the hematopoietic stem cells of mice. Strahlentherapie 140 (3): 302 .

Vasama R, Vasama R (1958). On the diurnal cycle of mitotic activity in the corneal epithelium of mice. Acta anat (Basel) 33: 230 .

Vinogradova G A (1961). The effect of the reparative regeneration of the liver on the mitotic activity of the corneal epithelium in mice. Bull Exp Biol Med (USSR) 50: 1198 .

CIRCADIAN HOST AND TUMOR RHYTHMS IN BALB/C MICE. RHYTHM INDUCTION IN HARDING-PASSEY MELANOMA

L. L. Sackett, E. Haus, D. Lakatua, J. Swoyer

Department of Anatomic and Clinical Pathology
St. Paul-Ramsey Medical Center
St. Paul MN 55101, U.S.A.

ABSTRACT

The Harding-Passey melanoma in Balb/C or CD_2F_1 female mice shows, in light synchronized (LD12:12) undisturbed animals, no detectable circadian rhythm in cell proliferation as gauged by ^{3}H-thymidine uptake in DNA. Manipulation of the animals and saline injection (.2 ml/20 gm), hydroxyurea injection (10 mg/.2 ml/20 gm), and ACTH-17 (HOE 433) (.4 IU/.2 ml/20 gm) induce a statistically significant circadian variation detected in several studies 4-24 and 16-36 hours after the treatment, only if the injection is given at the beginning of the light phase (LD12:12). Thus, tumor synchronization in this model is critically dependent upon the circadian stage of administration of the synchronizing agent. Endogenous and exogenous ACTH seem to be the synchronzing agent. Hydroxyurea in the dose given shows no additional effect.

Supported by the St. Paul-Ramsey Medical Education and Research Foundation (8256 and 8225) and USPH, National Cancer Institute (CA-14445).

INTRODUCTION

Mammalian cells do not divide and proliferate at random, but follow a rhythmic growth pattern in several frequency ranges. If one studies the number of mitoses or the ^{3}H-thymidine uptake during the S phase of the mitotic cycle by repeated sampling over a 24-hour span, one will almost invariably find that although some mitotic figures or some cells in S phase can be found at any time, the majority of cells in the growth fraction will form DNA and/or will undergo mitotic division at certain times which are recurring and predictable. The sequence and the time relation of the metabolic and proliferative circadian periodic events within the same tissue and within the organism as a whole are characteristic for its normal function. Different tissues in the mammalian body show differences in timing and, under conditions of health and environmental synchronization, firm time relations to each other which may be plus, minus or zero (Barnum, Jardetzky and Halberg, 1958; Edmunds, 1977; Levi, Halberg, Nesbit and others, 1980; Rensing and Goedeke, 1976; Scheving and Pauly, 1973).

If the events in a tissue which lead to cell division and proliferation follow a rhythmic pattern with certain (predictable) times in which statistically the majority of the dividing cells will be found in a certain stage of the cell cycle, the sensitivity of this tissue to agents which interfere with some aspect of cell metabolism or with one or the other stage of the cell cycle will also vary in a periodic and predictable fashion. Thus, the assumption appeared justified that the mammalian organism will show circadian periodic changes in sensitivity and resistance toward drugs which interfere with cell growth and cell multiplication. Empirically, this was verified for numerous physical and chemical agents (Halberg, Gupta, Haus and others, 1977; Haus, Halberg, Kuhl and Lakatua, 1974; Levi, Halberg, Nesbit and others, 1980; Scheving, Burns, Pauly and others, 1977; Scheving, von Mayersbach and Pauly, 1974). The latter include both cycle-specific and cycle-nonspecific chemotherapeutic agents.

The susceptibility-resistance cycles to agents used in cancer chemotherapy in non-tumor-bearing animals are thought to be due to the circadian periodicity of cell metabolism and/or proliferation in tissues critical for host tolerance and survival. Since this concept should equally apply to malignant growth, it appeared of interest to study the temporal behavior of cell proliferation in tumors. The results were quite variable with some spontaneous and experimental tumors showing a circadian periodicity in cell proliferation, while others did not allow the recognition of rhythms as group phenomenon of their proliferating units (Badran and Echave Llanos, 1965; Berezkin, 1970; Halberg, Gupta, Haus and others, 1977; Garcia-Sainz and Halberg, 1966; Levi, Halberg, Nesbit and others, 1980; Meng and Pohle, 1961).

In the design of timed, so-called "chronotherapeutic" treatment schedules, both host and tumor rhythms appear to be of interest. The most desirable situation for timed treatment would be if a chemotherapeutic agent could be administered at a time when the susceptible tissues of the host which limit the application or dosage of the agent are at the most resistant stage of their proliferative cycle, while the tumor is at the same time at its most sensitive one. If, however, host and tumor are synchronous and the most sensitive and the most resistant stage occur in both at the same time, no obvious benefit can be gained from timed treatment. In such a case, however, a phase shift of the host or of the tumor may be attempted. Situations have been described in which, by a change in synchronizer phase (i.e., the lighting regimen in rodents) host rhythms but not the tumor rhythms were phase-shifted (Haus and Halberg, 1978).

In other instances, the host tissues may be circadian periodic but the tumor may not. In such a case, treatment can be timed according to the susceptibility-resistance cycle of the host, with the drug given at the most resistant stage of the circadian cycle of the critical host tissues. Alternatively, it could be attempted to synchronize the tumor and/or to induce a circadian periodicity of proliferation in the malignant tissue different in phase from that of the critical host tissues. Attempts of tumor-cell synchronization have been made in experimental tumor models and in patients in order to allow tumor-oriented timed treatment (Ernst and Killman, 1971; Focan, LeHung, Derouaux and others, 1977; Klein, Adler, Doering and others, 1976; Klein, Lennartz, Habitch and others, 1970; Maidhof, Jellinghaus, Schultze and Maurer, 1975; Mauer, Murphy and Hayes, 1976; Pouillart, Schwarzenberg, Mathe and others, 1972; Rajewsky, 1970; Rajewsky, 1973). Some success has been reported with this approach in animal tumors (Klein, Adler, Doering and others,

1976; Klein, Lennartz, Habitch and others, 1970; Rajewsky, 1970) as well as in patients (Focan, 1975; Focan, LeHung, Derouaux and others, 1977; Klein, Adler, Doering and others, 1976; Lampkin, Nagao and Mauer, 1971; Lampkin, McWilliams and Mauer, 1971). Some investigators, however, reported a considerable variability (Klein, Adler, Doering and others, 1976; Lampkin, McWilliams, Mauer and others, 1976) in the results, lack of correlation of the clinical response with the cytokinetic findings (Jentsch, 1975; Vogler, Kremer, Knospe and others, 1976) or lack of therapeutic benefit in the experimental animals or patients (Ganzer, Ryzmann and Vosteen, 1977; Jellinghaus, Maidhof, Schulze and Maurer, 1975). It has to be emphasized, however, that these attempts of tumor synchronization were made merely with the intent to accumulate cells in a certain stage of the mitotic cycle without regard to the circadian stage of the host at the time of the treatment and without regard to possible pre-existing circadian rhythmicity of the tumor. Neglect of the time structure of host and tumor at the time of a chrono-therapeutic intervention may be one of the factors leading to conflicting results.

An experimental tumor used in this laboratory which did not show a circadian rhythm of ^{3}H-thymidine uptake when studied in undisturbed animals at six time points around the clock was the Harding-Passey melanoma. An attempt was made to synchronize this tumor, taking into account circadian system stage of the host.

MATERIALS AND METHODS

In five studies a total of 1,056 mice were used. For details see Table 1.

Harding-Passey melanoma was carried in Balb/c and CD_2F_1 female mice. A tumor-cell suspension in saline was prepared with a tissue press; and 0.1 ml of the suspension was injected subcutaneously in the right flank of the animals with the exception of the mice in Study I, which received their tumor by trocar. After injection of tumor, the mice were housed 10 per cage in plastic cages measuring 36 cm x 30 cm x 16 cm until 7-10 days prior to each study, at which time the animals were housed two per cage in plastic cages measuring 28 cm x 13 cm x 14 cm. Food and water were available ad libitum. The temperature was controlled by air conditioning

Table 1. Outline of five studies of tumor synchronization in mice carrying a Harding-Passey melanoma

Study	Date	Lighting Regimen	Strain/Sex	No. of Mice	Source	Age at Time of Killing (weeks)	Tumor Transplant* (days prior to killing)	Treatment Type	Treatment Time	Sacrifice Hours After R (first and last group)	Sacrifice 1st Time Pt.** Clock Hour	Sacrifice 1st Time Pt.** HALO***
1	Feb 1/2 1979	LD8:16 (L 08-16 D 16-08)	Balb/c Female	66	Laboratory Supply	11	20	0	--	--	12^{00}	04
				66				0	--	--		
2	Sept 25/28 1979	LD12:12 (L 06-18 D 18-06)	"	72	Microbiological Research, Inc.	17	29	OHU	$08^{00}, 20^{00}$	4 - 24	12^{00}	06
				72				S			00^{00}	18
3	Dec 19/21 1979	"	"	72	Simonsen laboratories	18	27	OHU	$08^{00}, 20^{00}$	16 - 36	00^{00}	18
				72				S			12^{00}	06
4	Apr 23/25 1980	"	"	84	"	36	21	OHU	$12^{00}, 00^{00}$	4 - 28	16^{00}	10
				84				S			04^{00}	22
5	Jun 17/19 1980	"	$CD_2 F_1$ Female	36	"	20	19	0	$08^{00}, 12^{00}, 16^{00}$ $20^{00}, 00^{00}, 04^{00}$	4 - 24	12^{00}	06
				216				S			16^{00}	10
				216				ACTH			20^{00}	14
											00^{00}	18
											04^{00}	22
			Total	1056							08^{00}	02

*Harding-Passey melanoma, 0.1 cc of cell suspension subcutaneously in right flank
**Sampling interval four hours for 24 hours (6 time points)
***HALO, hours after lights on

and the lighting regimen by an automatic light switch as
indicated in Table 1. The mice were kept on this regimen for
at least 21 days. During one week prior to each study the
rooms were entered only once a day quietly for checking of
the water bottles. The cages were changed once a week with
as little handling of the animals as feasible.

Study 1

In Study 1 the circadian rhythms in host and tumor were
explored. The mice were removed in their cage from the
animal room as quietly as possible. During the dark span
the working area was kept in dim red light. The animals
were removed singly from their cage, weighed, and their
rectal temperature was measured as a circadian periodic
reference function with a Digitec Thermistor thermometer
model #5810 with probe #702. They were then injected with
^{3}H-thymidine in saline (5 mcC/0.2 ml/20 gm body weight),
earmarked and placed in a plastic cage of the same dimension
and returned to the animal room. After 20 minutes, the
animals were returned to the laboratory and rapidly killed
by decapitation. Blood was collected for a white blood cell
count and a smear was prepared for a differential count.
The remainder of the blood was drained into a centrifuge
tube for the determination of corticosterone as reference
function. The tumor and a bone marrow sample (lumbar spine)
were removed and preserved in formalin.

Studies 2, 3 and 4

The effect on the ^{3}H-thymidine uptake in the tumor of
an injection of hydroxyurea (10 mg/0.2 ml/20 gm) or of
saline (0.2 ml/20 gm) given intraperitoneally to different
subgroups of mice 12 hours apart was investigated. In
Studies 2 and 3, hydroxyurea (OHU) or saline (S) were given
at 08^{00} and at 20^{00}. Under a lighting regimen of LD12:12
with light from 06^{00} to 18^{00} alternating with darkness, this
corresponds to a time of two Hours After Light On (HALO) and
14 HALO if the onset of the light span is taken as the phase
reference. In Study 4, OHU or S were given at 12^{00} (6 HALO)
or at 00^{00} (18 HALO). ^{3}H-thymidine injection, followed 20
minutes later by killing of subgroups of animals of both
treatment groups and by sampling such as outlined
for Study 1, was begun four hours after the injection

of OHU or S and continued at four-hourly intervals until 24 hours after injection in Study 2, and until 28 hours after injection in Study 4. In Study 3, ^{3}H-thymidine injection followed by sampling was begun at 16 hours after injection of OHU or S and continued at four-hourly intervals until 36 hours after injection.

Study 5

In Study 5, some animals received ^{3}H-thymidine without pretreatment at six time points ($\Delta t = 4$ hr), beginning at 12^{00} (06 HALO) and were killed and sampled 20 minutes later, as in the previous studies. Other groups of mice received either an intraperitoneal injection of saline (0.2 ml/20 gm body weight) or of a synthetic short-chain ACTH-17 (HOE 433) (0.4 IU/0.2 ml/20 gm body weight) at either one of six time points (08^{00}, 12^{00}, 16^{00}, 20^{00}, 00^{00} or 04^{00}). ^{3}H-thymidine injection (5 mcC/0.2 ml/20 gm body weight), followed in 20 minutes by killing and sampling, was begun four hours after saline or ACTH and continued in subgroups of each of these groups at four-hourly intervals for 24 hours. The parameters studied were the same as in the previous experiments.

For the determination of the incorporation of ^{3}H-thymidine into the DNA in bone marrow (femur) and tumor, the tissues were removed from the formalin fixative and cleaned from fibrous tissue, muscle and adipose tissue as far as possible.

^{3}H-thymidine incorporation into DNA was determined by the method of Ogur and Rosen (1970). The tissue was digested with 1N NaOH (tumor) or 0.3N KOH (bone marrow) overnight, and then neutralized with 1N HCl. RNA was removed with perchloric acid. The pellet was re-extracted with perchloric acid following 30 min. incubation at 70°C. Two ml of the supernate were transferred into 15 ml scintillation fluid (formula 950-A) and counted in a liquid scintillation counter (Packard 2650). Another 0.5 ml of the supernate was diluted with 0.5N perchloric acid and DNA content determined at 260 nm, using a Beckmann spectrophotometer. ^{3}H-thymidine uptake was expressed in DPM/mcg DNA.

The results were transferred to punch cards and analyzed by means of an electronic computer (Wang 2200). After inspection of the chronogram, possible circadian rhythms were assessed by single cosinor procedure (Halberg, Tong and Johnson, 1967), which yielded information on statistical rhythm detection and, if a rhythm could be demonstrated on the rhythm parameters: mesor (M), amplitude (A) and acrophase (ϕ) and their variance estimates.

RESULTS

A circadian rhythm in the incorporation of ^{3}H-thymidine into bone marrow DNA confirmed findings in Study 1 (Fig. 1). No such rhythm as a group phenomenon was detected in the Harding-Passey melanoma.

Studies 2, 3 and 4 are summarized in Figs. 2 and 3. After injection of saline (S) as well as of hydroxyurea (OHU) at the beginning of the light span (02 HALO), a circadian variation of ^{3}H-thymidine uptake in the DNA of the Harding-Passey melanoma comes to the fore which can be statistically validated by cosinor analysis both for the time span between 4 and 24 hours after injection (Study 2), and for the time span 16 to 36 hours after injection (Study 3) (Fig. 4). The circadian acrophase for the S- and for the OHU-injected animals is identical. The acrophase obtained by sampling 4-24 hours after S or OHU and that obtained by sampling 16-36 hours after the injection is found for the tumors of the S- and for the OHU-injected mice during the first half of the dark span. During the time span 16-36 hours after injection, the acrophase appears to be about $3\frac{1}{2}$ hours earlier than during the span 4-24 hours after injection. This observation will have to be confirmed and extended by following the tumor beyond the 36-hour span studied.

Injection of S as well as of OHU during the middle of the light span (06 HALO) and during the second half of the dark span leads to apparent changes in ^{3}H-thymidine uptake in tumor DNA which, however, during the time span studied do not allow a statistically significant rhythm detection. After injection two hours after the middle of the dark span (14 HALO), no appreciable change in ^{3}H-thymidine uptake occurred--the values obtained for S or OHU being not significantly different from those obtained without any treatment (Fig. 1).

In Study 5 a circadian rhythm in ^{3}H-thymidine uptake in DNA of Harding-Passey melanoma was obtained only after

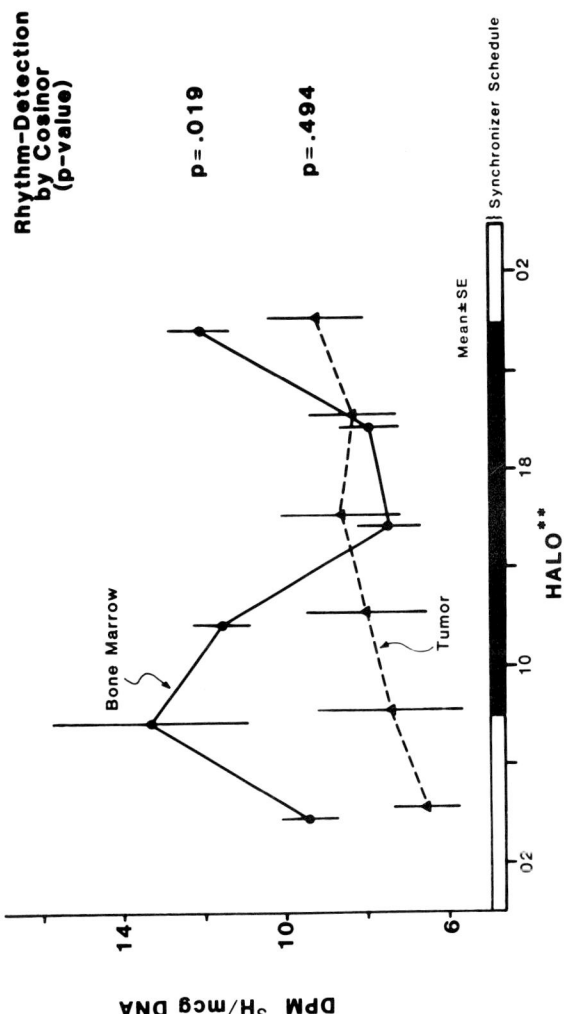

Fig. 1. Circadian rhythm in ^{3}H-thymidine uptake in DNA of Balb/c female mouse bone marrow. Absence of detectable rhythm in Harding-Passey melanoma (^{3}H-thymidine: 5 mcC/0.2 ml/ 20 gm body weight. i.p. 20 min. prior to killing. HALO = Hours After Light On.)

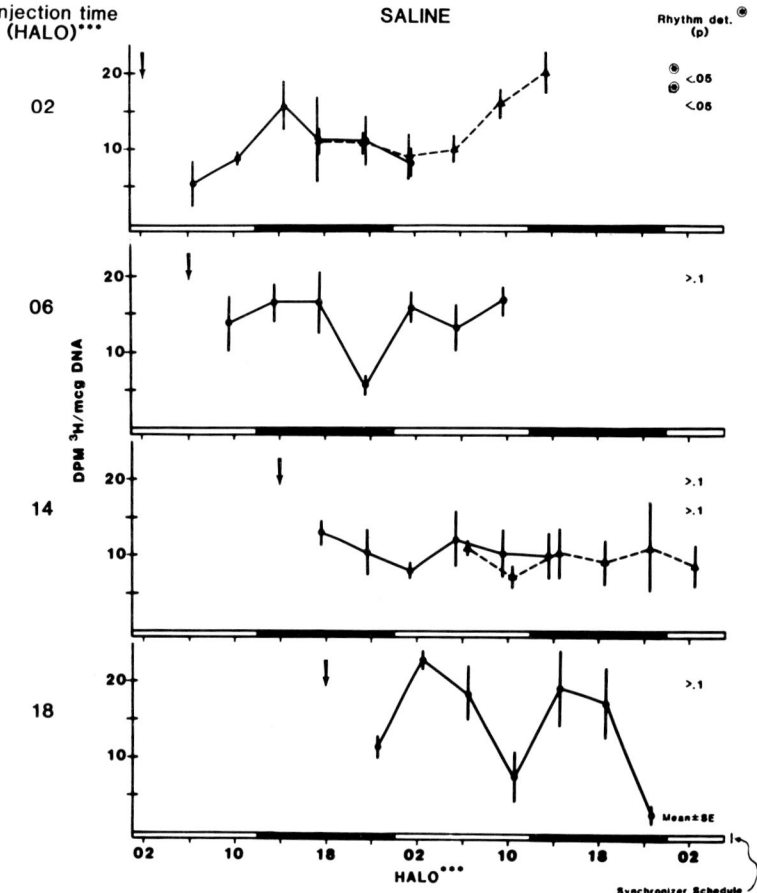

Fig. 2. Circadian-stage dependence of rhythm induction in Harding-Passey melanoma by handling and saline injection. ^{3}H-thymidine (5 mcC/0.2 ml/20 gm body weight) uptake in DNA of Harding-Passey melanoma in Balb/c female mice after saline (0.2 ml/20 gm body weight) injected at different circadian stages. Rhythm detection by single cosinor (p value). Sampling during hours after saline injection: ●——● 4-24; ▲---▲ 16-36.

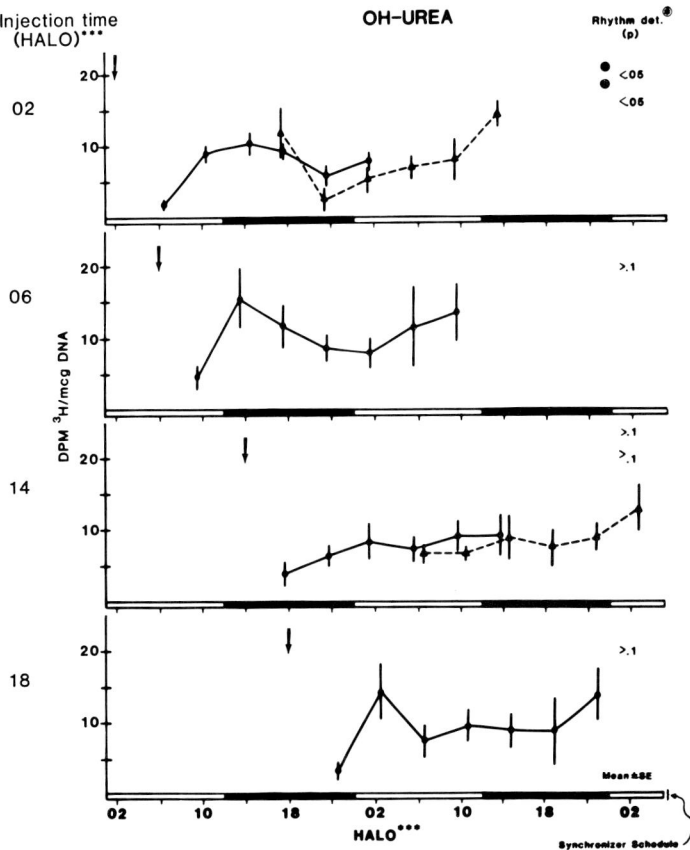

Fig. 3. Circadian-stage dependence of rhythm induction in Harding-Passey melanoma after hydroxyurea injection (10 mg/0.2 ml/20 gm body weight). Rhythm induction similar to that seen after saline alone. ^{3}H-thymidine (5 mcC/0.2 ml/20 gm body weight). Rhythm detection by single cosinor (p-value). Sampling during hours after OHU injection: ●——● 4-24; ▲--▲ 16-36.

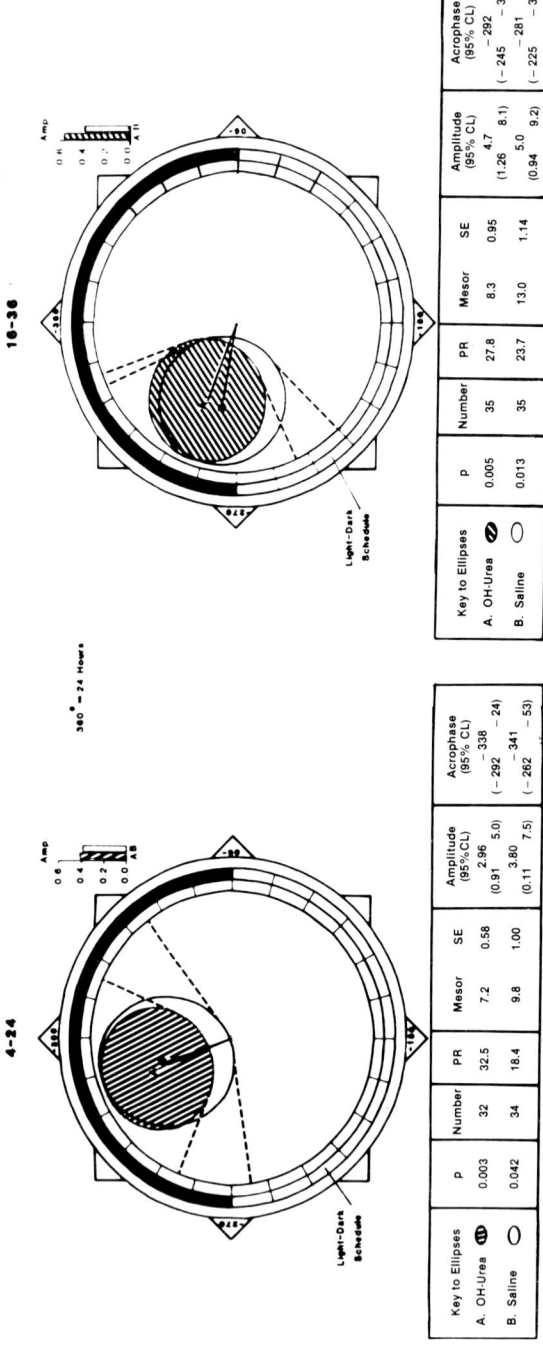

Fig. 4. Cosinor evaluation of circadian rhythm in ^{3}H-thymidine (5 mcC/0.2 ml/20 gm body weight) uptake in Harding-Passey melanoma in Balb/c female mice induced by injection of saline (0.2 ml/20 gm) or hydroxyurea (10 mg/0.2 ml/20 gm) at 02 HALO (Hours After Light On). Rhythm detection after both forms of treatment in tumors sampled 4-24 and 16-36 hours after treatment.

ACTH-17 (HOE 433) was given at the beginning of the light span (02 HALO). Injection of ACTH at the other time-points did not lead to a detectable circadian variation 4-24 hours thereafter. A circadian cycle in the S-treated animals was suggested only at the end of the dark span (22 HALO, an adjacent time point), and none was seen in the untreated mice. Figure 5 shows the ^{3}H-thymidine uptake in the untreated animals and in the mice which had received S or ACTH at 02 HALO. As indicated by the cosinor analysis in the left portion of the figure, a circadian variation was verified only for the ACTH-injected animals.

DISCUSSION

The time structure of tumors displaying rhythms of cell proliferation in different frequency ranges is of therapeutic interest. The detection of circadian as well as of circaseptan susceptibility resistance cycles of critical host tissues has led to the design of treatment schedules which achieved in several experimental models a marked decrease in toxicity of tumor chemotherapy. Because in clinical cancer chemotherapy, toxic doses of some therapeutic agents also have to be used, any improvement of the toxic-therapeutic ratio, either by the use of host rhythms or of tumor rhythms, may be very beneficial. The exploitation of the circadian susceptibility cycles may allow treatment at a time when the tumor is sensitive to the agent used--while the host is in a resistant stage of the proliferation cycle of his bone marrow or other critical tissues.

Tumor rhythms in patients are difficult to detect and to monitor. Synchronization by hormonal agents to which the tumor may be responsive or by pharmacologic agents which, e.g., delay passage through one or the other phase of the cell cycle, may help to overcome this problem. In any attempt of synchronization of a rhythmic biologic system including a tumor, pre-existing rhythms will have to be taken into account.

We tried to take advantage of circadian periodicity in host and tumor. We expected that the attempt of synchronization might yield different results at different circadian-system stages. Hydroxyurea was chosen as "synchronizing" agent on the basis of reports on tumor synchronization observed with this agent in other experimental tumor systems

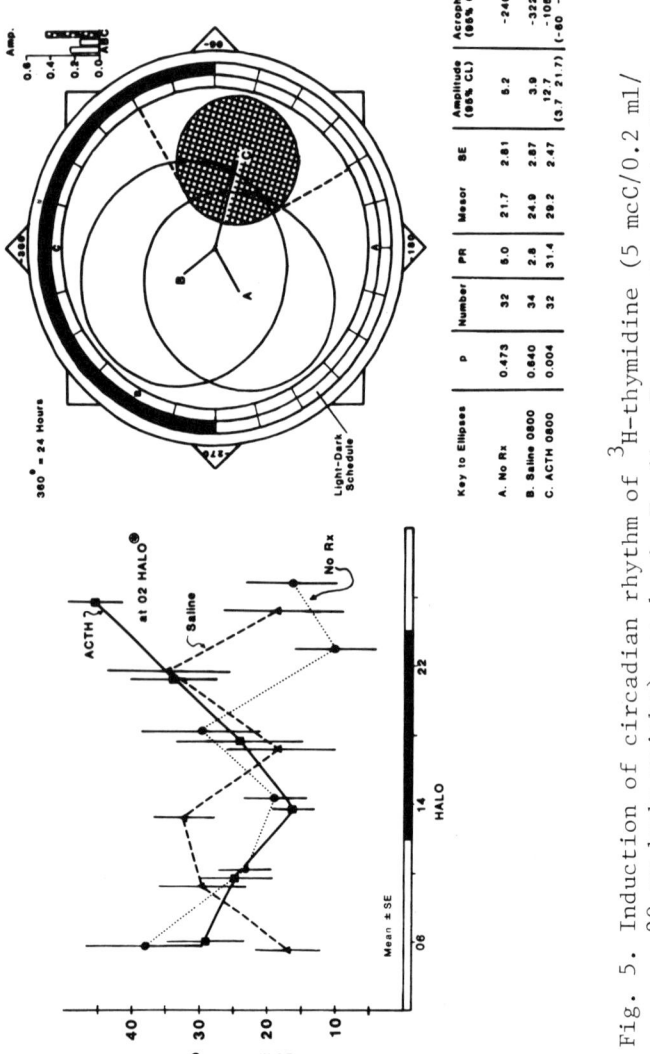

Fig. 5. Induction of circadian rhythm of ^{3}H-thymidine (5 mcC/0.2 ml/ 20 gm body weight) uptake in Harding-Passey melanoma in CD_2F_1 female mice by ACTH-17 (HOE 433) (0.4 IU ACTH-17/0.2 ml/ 20 gm body weight) only if given at 02 HALO. Same dose at 06, 10, 14, 18 and 22 HALO leads to no detectable circadian variation of ^{3}H-thymidine uptake in tumor.

(Rajewsky, 1973). We used saline injections to control for the stimulation provided by the process of handling and injection, and found that saline given shortly after onset of the light span synchronized the tumor just as much as hydroxyurea. Interpreting the saline effect as a manifestation of the host's response to the procedure of handling and injection, we postulated that this time-dependent effect may be due to the release of endogenous ACTH and presumably steroid hormones. In this case, the difference in timing might not be due to the difference in sensitivity of the tumor to the hormone, but to the sensitivity cycle of the pituitary in release of endogenous ACTH, which had been previously demonstrated (Haus, Halberg, Kuhl and Lakatua, 1974; Haus, 1964). This possibility, however, was made less probable by the demonstration of the same timing of rhythm induction by a fixed dose of exogenous ACTH in a dose which, by previous experimentation, had been shown to be close to maximal in its steroidogenic response (Haus, Halberg, Sothern and others, 1980; Brown, Halberg, Haus and others, 1980).

The induction of a circadian rhythm in cell proliferation in the Harding-Passey melanoma was limited to stimulation or to ACTH given early during the light span. The same procedure 12 hours later (early during the dark span) led to no change in ^{3}H-thymidine uptake measured 24 or 36 hours after the injection. At other time points such as 06 HALO and 18 HALO, some changes in the thymidine uptake were seen within 24 hours after injection. These, however, were irregular, and the recognition of a circadian rhythm by inferential statistical procedures was not possible.

This circadian-system phase dependence of the capability of a hormonal agent to induce a tissue to take up ^{3}H-thymidine in a circadian-periodic fashion over at least 36 hours after injection is of interest and has to be considered in all experimental designs aimed at synchronization of healthy or tumorous tissues.

The question arises whether the circadian rhythm manifested in the melanoma after injection of ACTH at 02 HALO represents the induction of a circadian rhythm or rather the synchronization of circadian-periodic elements which are out of phase with each other and therefore do not allow the recognition of a rhythm as a group phenomenon. Questions which will have to be answered are whether this rhythm is actually 24-hour periodic or whether it follows another circadian frequency, free-running from the host rhythms, and how long a rhythm induced or synchronized by ACTH in this tumor will continue? Also, it appears of interest whether host or tumor factors determine the apparent sensitivity cycle observed in the melanoma.

The studies reported show the feasibility of achieving tumor synchronization. It will have to be attempted to utilize this observation for the timing of chemotherapy. It appears that in the design of chemotherapeutic schedules, taking into account predictable changes in resistance and/or susceptibility, both of host and tumor together with rhythm manipulation, may lead to an improvement in the toxic/therapeutic ratio of cancer chemotherapy. Under certain circumstances, the advantages gained by a chronotherapeutically adjusted treatment schedule might make the difference between survival of the host or of the tumor from a given dose.

REFERENCES

Badran AF, Echave Llanos JM (1965). Persistence of mitotic circadian rhythm of a transplantable mammary carcinoma after 35 generations: its bearing on the success of treatment with endoxan. JNCI 2:285.

Barnum CP, Jardetzky CD, Halberg F (1958). Time relations among metabolic and morphologic 24-hour changes in mouse liver. Am J Physiol 195:301.

Berezkin MV (1970). A comparative study of the circadian rhythms of mitotic activity in neoplastic and normal tissues (Russian). Byull eksp Biol Med 70:83.

Brown H, Halberg F, Haus E, Lakatua D, Berg H, Sackett L, Melby J, Wilson T (1980). Circadian-stage-specified effects of a synthetic short chain ACTH-1-17 (HOE 433) on blood leukocytes and corticosterone secretion in mice. Chronobiologia 7 No. 1.

Edmunds LN (1977). Clocked cell cycle clocks: I. Cell cycle clocks, cell division cycles and circadian rhythms. II. Implications toward chronopharmacology and aging. In Proc Conf Biological Rhythms and Aging. St. Petersburg Beach, Florida. Waking and Sleeping 1:227.

Ernst P, Killman SA (1971). Perturbation of generation cycle of human leukemic myeloblasts in vivo by methotrexate. Blood 38:689.

Focan C, LeHung S, Derouaux M, Bury J, Noel JF, Claessens JJ (1977). Chimiotherapie sequentielle a visée de synchronisation-recrutement et rhythme circadien daus les tumeurs solides de l'adult: Bases theoriques et resultats d'un essai randomisé. Rev med de Liege 32:635.

Focan F (1975). Chimiotherapie sequentielle basée sur l'hypothese d'un rhythme circadien de la proliferation tumorale. Nouv Presse Med 4:2709.

Ganzer V, Ryzmann G, Vosteen KH (1977). Kritische Uberlegungen zur Zeitfolge der einzelnen Behandlungschritte bei der Synchronizationstherapie bösartiger Geschwülste. Arch Oto Rhino-Laryngology 214:291.

Garcia-Sainz M, Halberg F (1966). Mitotic rhythms in human cancer, re-evaluated by electronic computer programs--evidence for temporal pathology. J Nat Cancer Inst 37: 279.

Halberg F, Gupta BD, Haus E, Halberg E, Deka AC, Nelson W, Sothern RB, Cornelissen G, Klee J, Lakatua DJ, Scheving LE, Burns ER (1977). Steps toward a cancer chronopolytherapy. XIV Intl Cong of Therapeutics, Montpellier, France. L'Expansion Scientifique Francaise, Publ.

Halberg F, Tong YL, Johnson EA (1967). Circadian system phase--an aspect of temporal morphology; procedures and illustrative examples. In von Mayersbach H (ed): "The Cellular Aspects of Biorhythms," Berlin, Springer, p 20.

Haus E (1964). Periodicity in response and susceptibility to environmental stimuli. Ann NY Acad Sci 117:281.

Haus E, Halberg F (1981). Endocrine rhythms. In Scheving LH, Halberg F (eds): "Chronobiology--Principles and Application to Shifts and Schedules," NATO Advanced Study Inst., Series D, Sijthoff Intl Publ, Leiden, Netherlands.

Haus E, Halberg F (1978). Chronofarmacologia della neoplasia con speciale riferimento alla leucemia. In Bertelli A (ed): "Farmacologia Clinica e Terapia," CG Edizione Medico-Scientifiche, Turin, Italy, p 29.

Haus E, Halberg F, Loken MK (1974). Circadian susceptibility-resistance cycle of bone marrow cells to whole body X-irradiation in Balb/c mice. In Scheving LE, Halberg F, Pauly JE (eds): "Chronobiology," Igaku Shoin, Ltd. Tokyo, p 115.

Haus E, Halberg F, Scheving LE, Simpson H (1979). Chronotherapy of cancer--a critical evaluation. Intl J of Chronobiol 6:67.

Haus E, Halberg F, Sothern RB, Lakatua K, Scheving LE, de la Pena SS, Sanchez E, Melby J, Wilson T, Brown H, Berg H, Levi F, Culley D, Halberg E, Hrushesky W, Pauly J (1980). Time-varying effects in mice and rats of several synthetic ACTH preparations. Chronobiologia 7: no. 2.

Haus E, Halberg F, Kuhl JF, Lakatua DL (1974). Chronopharmacology in animals. In Aschoff J, Ceresa F, Halberg F (eds): "Chronobiological Aspects of Endocrinology," FK Schattauer Verlag, Stuttgart, New York. Chronobiologia 1:suppl 1, p 122.

Jellinghaus W, Maidhof R, Schulze B, Maurer W (1975). Experimentelle Untersuchungen und zellkinetische Beobachtungen zur Frage der Synchronization mit Vincristin in vivo (Mauseleukamie L1210, Krypten Epithelien der Maus), Z Krebsforsch 84:161.

Jentsch K (1975). Die Wirkung von 5-Fluorouuazil und Bestrahlung auf die Zellkinetik des Walker-Karzinoms und Dünndarms von Ratten. Strahlentherapie 150:51.

Klein HL, Adler D, Doering M, Klein PJ, Lennartz KJ (1976). Investigations on pharmacologic induction of partial synchronization of tumor cell proliferation: its relevance for cytostatic therapy. Cancer Treatment Rpts 60:1959.

Klein HO, Lennartz KJ, Habitch W, Eder M, Gross R (1970). Synchronisation von Ehrlich-aszites Tumorzellen und ihre Bedentung bei der Anwendung eines alkylierenden cytostaticum. Klin Wschr 48:1001.

Lampkin BC, McWilliams NB, Mauer AM (1971). The advantage of cell synchronization in the therapy of myeloid leukemias in children. Blood 38:802.

Lampkin BC, McWilliams NB, Mauer AM, Flessa HC, Hake DA, Fisher V (1976). Manipulation of the mitotic cycle in the treatment of acute myelogenous leukemia. Brit J Haematol 32:29.

Lampkin BC, Nagao T, Mauer AM (1971). Synchronization and recruitment in acute leukemia. J Clin Invest 50:2204.

Levi F, Halberg F, Haus E, de la Pena SS, Sothern RB, Halberg E, Hrushesky W, Brown H, Scheving LE, Kennedy BJ (1980). Synthetic adrenocorticotropin for optimizing murine circadian chronotolerance for adriamycin. Chronobiologia 7: no 2.

Levi F, Halberg F, Nesbit M, Haus E, Levine E (1980). Chrono-oncology. Submitted for publication.

Maidhof R, Jellinghaus W, Schultz B, Maurer W (1975). Experimentelle und theoretische Untersuchungen zur Erzeugung einer teilsynchron proliferierenden Zellpopulation mit Vincristin in vivo. Dtsch Med Wschr 100:54.

Mauer AM, Murphy SB, Hayes FA (1976). Evidence for recruitment and synchronization in leukemia and solid tumors. Cancer Treatment Rpt. 60:1841.

Meng K, Pohle K (1961). Die 24-Stunden Mitose-Rhythmik beim Ehrlichschen Mause - Ascites - Carcinom nach Ovarektomie und Adrenalektomie. Z Krebsforsch 64:219.

Nelson W, Zinneman H, Halberg F, Bazin H (1974). Circadian rhythm in Bence-Jones protein excretion by LOU rat bearing a transplantable immunocytoma, responsive to adriamycin treatment. Int J Chronobiol 2:327.

Ogur M, Rosen G (1970). The nucleic acids of plant tissues. I. The extraction and estimation of desoxypentose nucleic acid and pentose nucleic acid. Arch Biochem 25:262.

Pouillart P, Schwarzenberg L, Mathe G, Schneider M, Jasmin C, Hayat M, Weiner R, DeVassal F, Amiel JL, Beyer HP, Fajbisowitz S (1972). Essai clinique de combinaisons chimiothérapiques basées sur la notion de tentative de synchronization cellulaire. Nouv Presse Med I:1757.

Rajewsky MF (1973). Cinetique de proliferation des populations cellulaires et traitement des cancers. Bull du Cancer 60:143.

Rajewsky MF (1970). Synchronization in vivo: kinetics of a malignant cell system following temporary inhibition of DNA synthesis with hydroxyurea. Exp Cell Res 26:269.

Rensing L, Goedeke K (1976). Circadian rhythm and cell cycle possible entraining mechanisms. Chronobiologia 3:53.

Scheving LE, Burns ER, Pauly JE (1977). Chemotherapy of L1210 leukemia with cyclophosphamide, vincristine, cis-platinum diammine dichlorosis and methylprednisolone. Chronobiologia 4:178.

Scheving LE, Burns ER, Pauly JE, Halberg F, Haus E (1977). Survival and cure of leukemic mice after circadian optimization of treatment with cyclophosphamide and arabinosyl cytosine. Cancer Res 37:3648.

Scheving LE, Pauly JE (1973). Cellular mechanisms involving biorhythms with emphasis on those rhythms associated with the S and M stages of the cell cycle. Int J Chronobiol 1:269.

Scheving LE, von Mayersbach H, Pauly JE (1974). An overview of chronopharmacology. J Europ Toxicol 7:203.

Vogler WR, Kremer WB, Knospe WH, Omura GA, Tornyos K (1976). Synchronization with phase specific agents in leukemia and correlation with clinical responses to chemotherapy. Cancer Treatment Rpts 60:1845.

CIRCANNUAL RHYTHMS

NYCTOHEMERAL AND SEASONAL VARIATIONS IN THE NUMBER OF TRITIATED THYMIDINE LABELLED CELLS IN THE EPIPHYSEAL CARTILAGE OF THE TIBIA IN THE GROWING RAT. EFFECT OF LIGHTING DURATION AND TEMPERATURE

C. Oudet and A. Petrovic

F.R.A. 15, INSERM, Institut de Physiologie, Faculté de Médecine; 4, rue Kirschleger 67085 Strasbourg Cedex, France

Cartilages either originate from the cartilaginous primordium or constitute after the beginning of ossification. The former ones are called primary, the latter, secondary cartilages.

To the first category belong the epiphyseal cartilages of the long bones; to the second one belong, e.g., the condylar and coronoid cartilages of the mandible, the cartilage of the midpalatal suture, and the cartilage formed, in some cases, in the fracture callus (Petrovic and Stutzmann, 1979). In the last years, a series of experiments designed to detect behavioral differences between the two types of cartilages was performed. It could be demonstrated that, if the growth rate of all cartilages is dependent on general hormonal and nutritional factors (STH, somatomedin, testosterone, thyroxin...) only the growth rate of secondary cartilages could be modulated by local biomechanical manipulations (Charlier et al., 1968; Charlier et al., 1969; Petrovic et al., 1975; Stutzmann, 1976); primary cartilages do not respond to such manipulations by modifying their growth rate. The existence of regulating or servo-systems, including numerous relays governing the growth rate of secondary cartilages, was demonstrated (condylar cartilage, midpalatal suture,....) (Petrovic et al., 1975; Stutzmann and Petrovic, 1976). Moreover, the existence of seasonal and nyctohemeral variations in the growth rate of the condylar cartilage of the mandible was detected (Oudet, 1979; Oudet and Petrovic, 1977a,b; 1978a,b). Furthermore, the responsiveness of the condylar cartilage to biomechanical manipulations was shown to exhibit similar variations.

Only partial information on nyctohemeral variations and none on seasonal variations in epiphyseal cartilage are available (Simmons, 1962, 1964, 1968, 1974). Growth processes of primary cartilages are not subjected to similar regulating systems; in this respect, growth phenomena in secondary and primary cartilages differ from each other. It seemed thus appropriate to investigate whether environmental factors like lighting duration and temperature, which modify the growth rate of the condylar cartilage, could also have some influences on the growth rate of primary cartilages. Accordingly, we looked for the possible response of the distal epiphyseal cartilage of the tibia in the young rat to the aforementioned environmental factors at different seasons and times of the day.

In March and November, after having grown until their 28th day of life under ordinary conditions of lighting and temperature, Sprague Dawley male rats were housed for 4 weeks in controlled rooms. Additionally, in November, three lighting durations (8 hours, 12 hours and 16 hours) and three temperatures (14°C, 19°C and 27°C) were imposed.

In each experiment, rats were injected one hour before sacrifice with tritiated thymidine. Sixteen rats were killed at each time point, i.e., 0100, 0700, 1300 and 1900. The tibias were removed, fixed, serially sectioned for radioautography and staining with toluidin blue. The total number of labelled cells was counted on every tenth section in each distal epiphyseal cartilage; this value is an operational estimation of the growth rate of the cartilage.

In March and November, when rats had been submitted for 4 weeks to 12 hours of light and 12 hours of darkness at a constant 19°C temperature, the number of labelled chondroblasts in the epiphyseal cartilage exhibited highly significant nyctohemeral variations ($p<0.001$) with a peak around 1300. Seasonal variations in the growth rate (Fig. 1) are also highly significant ($p<0.001$); the number of labelled chondroblasts is higher in March than in November. However, there are no detectable interactions between the effects of the nyctohemeral and seasonal growth rhythms. The pattern of the nyctohemeral variations is asymmetric: the maximum of labelled cells is about 20% higher than the mean level, while the minimum is only about 10% lower than the mean level (Fig. 2).

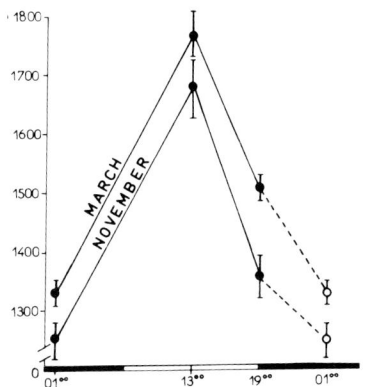

Fig. 1

Nyctohemeral variations in the number of tritiated thymidine-labelled cells in the epiphyseal cartilage of the tibia of 56 day-old rats, in March and November. Each point is the mean for 16 observations, and the vertical line its standard error.
Light period ═══
Dark period ▬▬▬

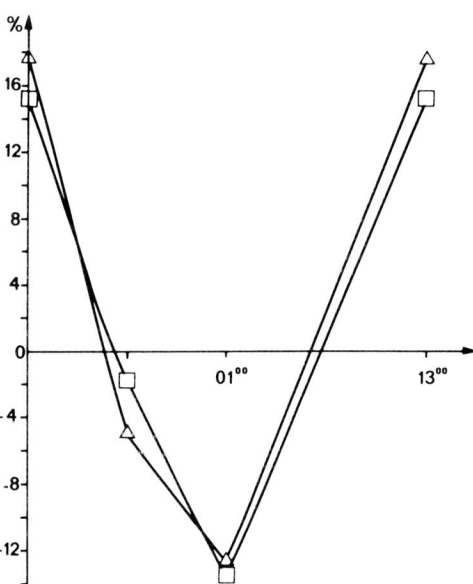

Fig. 2
Pattern of nyctohemeral variations of the number of tritiated thymidine-labelled cells in the epiphyseal cartilage in March and November. The origin of the axis is the middle of the rest period of the rat. Variations are expressed as % of the means.

In November, after the rats had been subjected to lighting durations of either 8, 12 or 16 hours since 4 weeks of age, the middle of the light period corresponded to the middle of the rest period of the rat, i.e., 1300. Whatever the duration of lighting, the growth rate of the epiphyseal cartilage exhibits highly significant nyctohemeral variations (Fig. 3) ($p<0.001$), with a peak around 1300. The number of labelled chondroblasts was the highest when the lighting duration was 16 hours ($p<0.001$) and the lowest when it was 8 hours. No interactions between the effects of lighting duration and the nyctohemeral rhythm of growth in the cartilage could be detected. Amplitude and acrophases (calculated and depicted according to Halberg method) (Halberg et al., 1972) of nyctohemeral variations under the three different lighting durations are not significantly different (Fig. 4).

In November, when rats had been subjected since 4 weeks of age to various environmental temperatures, i.e., 14°C, 19°C or 27°C, the fastest growth rate of the epiphyseal cartilage was observed at 19°C and the slowest one at 27°C. For each of the three environmental temperatures, the number of labelled chondroblasts, however, exhibits highly significant nyctohemeral variations; the acrophases and amplitudes of these daily rhythms are of the same order. No significant interactions between the effects of temperature and the effects of daily variations of the number of labelled chondroblasts were detected (Figs. 5 and 6).

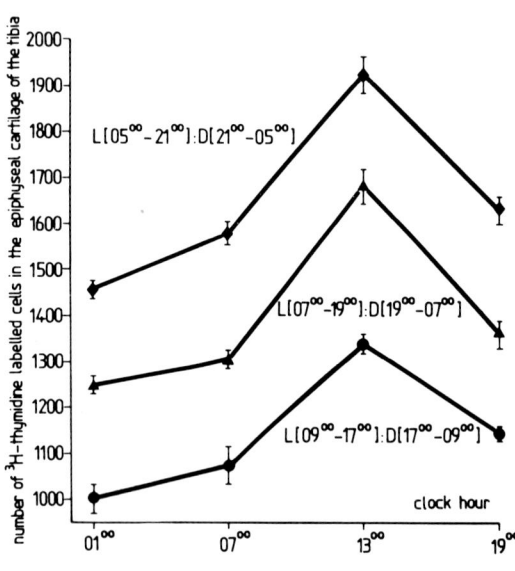

Fig. 3. Nyctohemeral variations of the number of tritiated thymidine labelled chondroblasts in the epiphyseal cartilage of the tibia of 56 day-old rats subjected to 3 different lighting durations since 4 weeks of age. Each point is the mean for 16 observations, and the vertical line is its standard error.

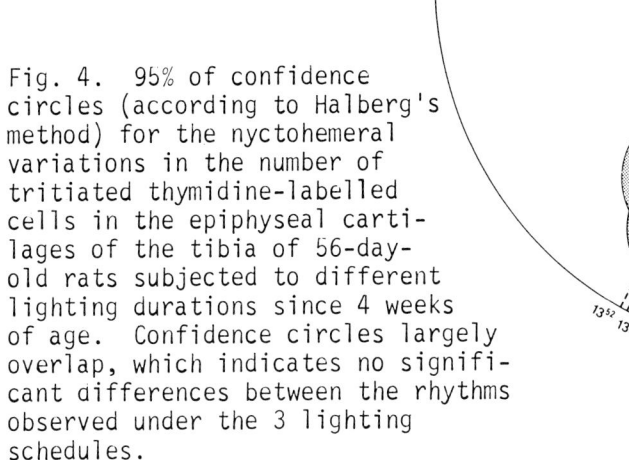

Fig. 4. 95% of confidence circles (according to Halberg's method) for the nyctohemeral variations in the number of tritiated thymidine-labelled cells in the epiphyseal cartilages of the tibia of 56-day-old rats subjected to different lighting durations since 4 weeks of age. Confidence circles largely overlap, which indicates no significant differences between the rhythms observed under the 3 lighting schedules.

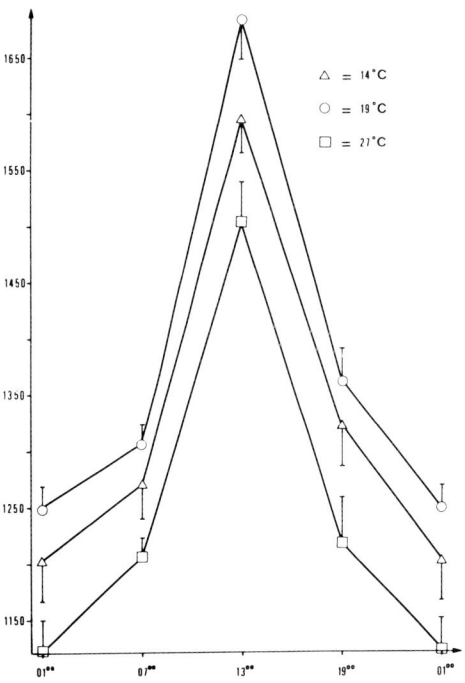

Fig. 5

Nyctohemeral variations in the number of tritiated thymidine-labelled chondroblasts in the epiphyseal cartilages of the tibia of 56 day-old rats subjected to 3 different temperatures since 4 weeks of age. Each point is the mean for 16 observations, and the vertical line is its standard error.

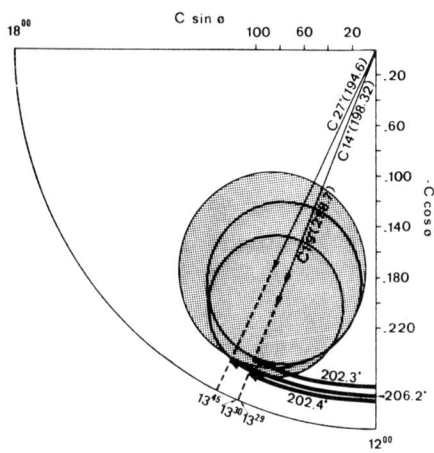

Fig. 6.

95% of confidence circles (according to Halberg's method) for the nyctohemeral variations in the number of tritiated thymidine-labelled chondroblasts in the epiphyseal cartilage of the tibia of 56-day-old rats subjected to different temperatures since 4 weeks of age. Confidence circles largely overlap, which indicates no significant differences between the rhythms observed under the 3 temperatures.

When one compares the behavior of a primary cartilage, the distal epiphyseal cartilage of the tibia, to the behavior of a secondary cartilage, the condylar cartilage of the mandible, with respect to imposed lighting duration and temperature, some similarities are apparent:

- longer the lighting duration, greater the growth rate in both cartilages;
- a high environmental temperature decreases the growth rate of both cartilages.

But, some differences have to be noted:

- the change of lighting duration (more precisely the change of the hours of beginning and end of the light period) causes a shift in the rhythm of growth rate of the condylar cartilages; we didn't detect such a phenomenon in the epiphyseal cartilage;
- the maximum of dividing cells was observed at a 14°C temperature in the condylar cartilage and at a 19°C temperature in the epiphyseal cartilage.

In the experiments reported above, the response to environmental factors was less complex in the epiphyseal cartilage than in the condylar cartilage. One possible explanation for this behavioral difference could be found in the histological structure: in the condylar cartilage, dividing cells were not yet surrounded with cartilaginous matrix; not unexpectedly, they were directly subjected to local constraining factors. In the epiphyseal cartilage, dividing cells were isolated by the cartilaginous matrix which seemed to "protect" them from local influences. Hence, perhaps they responded less to the environmental factors investigated in this report.

From an evolutionary standpoint, the adaptive features of the condylar-cartilage growth, by reason of facilitating the optimal occlusal adjustment, present an obvious selective advantage in the mandible; on the contrary, while largely independant of extrinsic local factors, the epiphyseal cartilage growth provides an obvious selective advantage in the limbs.

In conclusion, the epiphyseal cartilage of the tibia responds to environmental factors like lighting duration and temperature variations, by modifying its growth rate. But the response remains less complex than in a secondary cartilage, like the condylar cartilage.

REFERENCES

Charlier J-P, Petrovic A, Herrmann J (1968). Déterminisme de la croissance mandibulaire: effets de l'hyperpropulseur et de l'hormone somatotrope sur la croissance condylienne de jeunes rats. Orthod Fr 39:567.

Charlier J-P, Petrovic A, Herrmann-Stutzmann J (1969). Effects of mandibular hyperpropulsion on the perchondroblastic zone of young rat condyle. Am J Orthod 55:71.

Halberg F, Johnson EA, Nelson W, Runge W, Sothern R (1972). Autorhythmometry. Procedures for physiologic self-measurements and their analysis. Physiology Teacher 1:1.

Oudet C (1979). Rythme nycthéméral et saisonnier de la vitesse de croissance du squelette et de la susceptibilité du cartilage condylien á l'égard des dispositifs orthopédiques. Thése de Doctorat en Biologie Humaine, Université Louis Pasteur, Strasbourg, France, 224 p.

Oudet C, Petrovic A (1977a). Effects of a postural hyperpropulsor in the growth of the mandibular condylar cartilage of the young rat during circadian cycle. In Lassman G, Seitelberger F, eds: "Rhythmische Funktionen in Biologischen Systemen, Ihre Bedeutung für Theorie und Klinik," Facultas-Verlag, Wien, II. Teil, 173 p.

Oudet C, Petrovic A (1977b). Circannual growth variations of the mandibular condylar cartilage in the young rat. J Interdiscipl Cycle Res 8:338.

Oudet C, Petrovic A (1978a). Growth rhythms of the cartilage of the mandibular condyle. Effects of orthopaedic appliances. Int J Chronobiol 5:545.

Oudet C, Petrovic A (1978b). Growth rhythms of the cartilage of the mandibular condyle. Effects of an orthopaedic appliance. In Reinberg A, Halberg F, eds: "Chronopharmacology," Proceedings of Satellite Symposium of 7th International Congress of Pharmacology, Paris, 21-24/7 1978, Pergamon Press, 65.

Petrovic A, Stutzmann J (1979). Contrôle de la croissance post-natale du squelette facial. Données expérimentales et modéle cybernétique. Actualités Odonto-Stomatologiques 128:811.

Petrovic A, Stutzmann J, Oudet C (1975). Control processes in the postnatal growth of the condylar cartilage of the mandible. In McNamara JA, ed.: "Determinants of Mandibular Form and Growth," Monogr. 4, Cranio-Facial Growth Series, Center for Human Growth and Development, Univ. Michigan, Ann Arbor.

Simmons DJ (1962). Diurnal periodicity in epiphyseal growth cartilage. Nature 195:82.

Simmons DJ (1964). Circadian mitotic rhythm in epiphyseal cartilage. Nature 202:906.

Simmons DJ (1968). Daily rhythms of S^{32} incorporation into epiphyseal cartilage in mice. Experientia 24:363.

Simmons DJ (1974). Chronobiology of endochondral ossification. Chronobiologia 1:97.

Stutzmann J (1976). Particularités de la croissance post-natale des cartilages secondaires du squelette facial. Recherches in vivo et en culture organotypique, chez le jeune rat, sur les processus de commande et régulation. Thése de Doctorat d'Etat ès-Sciences, Université Louis Pasteur, Strasbourg, France, 230 p.

Stutzman J et Petrovic A (1976). Experimental analysis of general and local extrinsic mechanisms controlling upper jaw growth. In McNamara JA, ed.: "Factors Affecting the Growth of the Midface," Monogr. 6, Cranio-Facial Growth Series, Univ. Michigan, Ann Arbor, 399 p.

Grants C.R.L. INSERM No. 76.1.135.4
A.T.P. INSERM No. 42.76.74.13

SEASONAL VARIATIONS IN THE DIRECTION OF GROWTH OF THE MANDIBULAR CONDYLE

A. PETROVIC, J. STUTZMANN and C. OUDET

F.R.A. 15, INSERM, Institut de
Physiologie, Faculté de Médecine,
4, rue Kirschleger
67085 Strasbourg Cedex, France

In young growing rat, the sagittal adjustment of mandible to maxilla has been shown to result from extrinsic regulation processes organized as a servosystem (Petrovic, Stutzmann, Oudet, 1975 ; Petrovic, Stutzmann, 1977 ; Petrovic, 1977 ; Gasson, Stutzmann, Petrovic, 1975). This adjustment occurs through a change in growth rate of the condylar cartilage, as well as a change in growth direction of the condyle. Additionally, the growth rate of the condylar cartilage, phylogenetically and ontogenetically, a secondary cartilage, can be modulated not only through hormonal influences but also through the biomechanical action of orthopedic appliances (Charlier, Petrovic, Herrmann-Stutzmann, 1968 ; Charlier, Petrovic, Herrmann-Stutzmann, 1969 ; Petrovic, Stutzmann, Oudet, 1975). The biological relays of these appliances are muscles, and some other local structures. The variations of the direction of growth of the condyle involve variations in the sagittal distribution of the mitoses in the condylar cartilage, and the orientation of the newly formed endochondral bone trabeculae (Stutzmann, 1976).

The object of this investigation was to relate the variations of the growth direction of the condyle to the seasonal variations of the condylar cartilage growth rate. In this experiment, the direction of growth of the condyle has been investigated in May and in November, in young control rats and in rats treated for 4 weeks with an orthopedic appliance, the so-called postural hyperpropulsor (Fig. 1), known to increase the lengthening of the mandible.

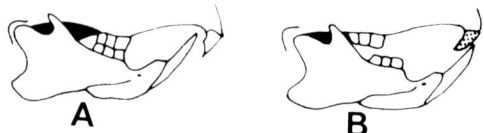

Fig. 1 - Postural hyperpropulsion of the mandible in the young rat.
A : Normal occlusal adjustment of the rat; in black, the lateral pterygoid muscle.
B : Occlusion after the setting of the appliance on the upper incisors ; the lower dental arch is put in a more forward position and the relationship between the two jaws are inverted.

The measure of the direction of growth of the condyle was done according to Stutzmann's method : one measures the angle between the centrally located, newly formed bone trabeculae and the mandibular plane (or any other reference concerning the corpus of the mandible, e.g. intraosseous metallic implants) (Fig. 2) (Stutzmann, 1976). Indeed, this method appears to be the most accurate, the most reliable, and the earliest parameter of variations in growth direction of the condyle.

In May and November, young Sprague-Dawley male rats were housed under ordinary conditions of lighting and temperature until their 28th day. When 4 week-old, they have been submitted to a 12-hours light and 12-hours darkness cycle at a constant 19° C environmental temperature. Half of them were treated during 10 hours a day, during the light period, with a "postural hyperpropulsor" of the mandible. All rats were killed at the age of 56 days. The mandibles were X-rayed, and the angle as defined above was measured.

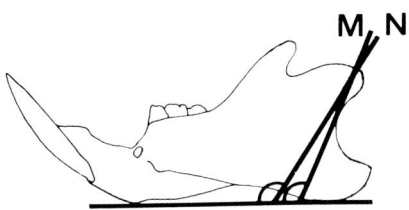

Fig. 2 - Measurement of the angle between the newly-formed endochondral bone trabeculae below the condylar cartilage and the mandibular plane of the young rat, in May (M) and November (N). The angle opens between May and November.

The measured angle is significantly smaller in May than in November (Fig. 3, Tab. 1).

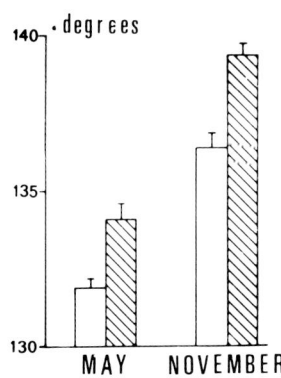

Fig. 3 - Means (columns) and standard errors (vertical lines) for 16 observations of the angle in control rats (white) and in rats treated since 4 weeks with a postural hyperpropulsor of the mandible (hachured), in May and in November.

Source of variation	Sum of squares	d.f.	Mean square	F
Month	190	1	190	128***
Hyperpropulsion	55	1	55	37**
Interaction	1	1	1	< 1 N.S.
Error	41.5	28	1.48	
Total	287.5			

Table 1 - Analysis of variance of the variations of the angle originating from the month and a postural hyperpropulsion treatment.

Both in May and in November, the wearing of the "postural hyperpulsor" leads to a significant increase in the angle (Fig. 3 Table 1).

No significant interaction between seasonal variations of the angle and the effects of the postural hyperpropulsion could be detected. The calculated value of the interaction is :

I = (131.9 + 139.4) - (134.1 + 136.4) = 0.8 (N.S.).

In other words, variations due to the season and variations due to the action of the orthopedic treatment are simply additive (Tab. 1).

It has previously been established that the growth rate of the condylar cartilage is greater in May than in November (Oudet, Petrovic, 1976 ; Oudet, Petrovic, 1977 ; Oudet, Petrovic, 1978), and that the "postural hyperpulsor" increases the growth rate of the condylar cartilage (Charlier, Petrovic, Herrmann-Stutzmann, 1968 ; Charlier, Petrovic, Herrmann-Stutzmann, 1969 ; Petrovic, Stutzmann, Oudet, 1975). But, if these two factors both accelerate the growth rate, their effects on the growth direction of the condyle are opposite. Indeed, when the acceleration of the growth rate results from the seasonal variations (in May), the supplement of mitoses is located, sagittaly, mostly

in the upper part of the condylar cartilage ; in
parallel, newly formed endochondral bone trabeculae
have a more vertical direction (the angle closes)
(Stutzmann, 1976). When the acceleration of the
growth rate is produced through the "postural hyperpropulsor", the supplement of mitoses is located,
sagittaly, mostly in the posterior part of the condylar cartilage ; in parallel, newly-formed endochondral bone trabeculae have a more posterior
direction (the angle opens)(Stutzmann, 1976).

How can we explain these results ?

The supplement in the lengthening produced by
b-STH* and its specific serum mediators is greater
in mandible than in the maxilla, when the incisors
have been resected (Charlier, Petrovic, Herrmann-
Stutzmann, 1968 ; Charlier, Petrovic, Herrmann-
Stutzmann, 1969 ; Petrovic, 1977). But, when the
incisors are present, and for the doses of b-STH in
physiological range, the lengthening is roughly the
same in the upper and the lower jaw. Indeed, the
incisors act as a "comparator" of the servosystem,
the upper ones being the "constantly changing reference input", and the lower ones being the "controlled variable" (Petrovic, 1977 ; Stutzmann, Petrovic,
1978). The tendency to a more forward position of
the lower incisors is detected as "deviation" from
the optimal occlusal adjustment ; the "deviation
signals" originating from the "detectors" of such a
deviation will bring about a decrease in the contractile activity of the lateral pterygoid muscle
detected electromyographically and, consequently, a
decrease in the iterative activity of the retrodiscal pad** (menisco-temporo-condylar fraenum),
resulting in a less important increased growth rate

---

* b-STH, National Pituitary Agency (U.S.A.)
** Previous experiments have demonstrated that the
retrodiscal pad exerts its action on the condyle
biomechanically and through its blood and lym-
phatic vessels supply (Petrovic, Stutzmann,
1977).

of the condylar cartilage and in a less posterior growth direction of the condyle. On the contrary, in the presence of the "postural hyperpropulsor", when closing the mouth, the animal has to put the mandible in a more forward position, implying an increased contractile activity of the lateral pterygoid muscle (and an increased iterative activity of the retrodiscal pad), and resulting in an actual increased growth rate of the condylar cartilage, a more posterior location of the supplementary mitoses, and in a more posterior growth direction of the condyle.

In conclusion, the direction of growth of the mandibular condyle exhibits seasonal variations. Reported experiments enable a better understanding of the involved biological phenomena, especially of those related to occlusal adjustment. The measurement of the "Stutzmann's angle" could be a valuable element of diagnosis and estimation of treatment effectiveness in dento-facial orthopedics.

REFERENCES

Charlier J-P, Petrovic A, Herrmann-Stutzmann J (1968). Déterminisme de la croissance mandibulaire : effets de l'hyperpropulsion et de l'hormone somatotrope sur la croissance condylienne de jeunes rats. Orthod Fr 39:567.

Charlier J-P, Petrovic A, Herrmann-Stutzmann (1969). Effects of mandibular hyperpropulsion on the prechondroblastic zone of young rat condyle. Am J Orthod 55:71.

Gasson N, Stutzmann J, Petrovic A (1975). Les mécanismes régulateurs de l'ajustement occlusal interviennent-ils dans le contrôle de la croissance du cartilage condylien ? Expériences d'administration d'hormone somatotrope et de résection du cartilage septal chez le jeune rat. Orthod Fr 46:77.

Oudet C, Petrovic A (1976). Effets d'un rétropulseur actif sur la vitesse de croissance du cartilage condylien du jeune rat au cours du nycthémère et de l'année. Signification de la direction de croissance condylienne. Orthod Fr 47:15.

Oudet C, Petrovic A (1977). Circannual growth variations of the mandibular condylar cartilage in the young rat. J interdiscipl Cycle Res 8:338.

Oudet C, Petrovic A (1978). Growth rhythms of the cartilage of the mandibular condyle. Effects of orthopedic appliances. Int J Chronobiol 5:545.

Petrovic A (1977). L'ajustement occlusal, son rôle dans les processus physiologiques de contrôle de la croissance du cartilage condylien. Orthod Fr 48:337.

Petrovic A, Stutzmann J (1977). Further investigations into the functioning of the "comparator" of the servosystem (respective positions of the upper and lower dental arches) in the control of the condylar cartilage growth rate and of the lengthening of the jaw. In : The Biology of Occlusal Development, Monograph N° 6, J.A. Mc Namara ed., Cranio-Facial Growth Series, Center for Human Growth and Development, Univ. Michigan, Ann Arbor, 331 p.

Petrovic A, Stutzmann J, Oudet C (1975). Control processes in postnatal growth of condylar cartilage of the mandible. In : Determinants of Mandibular Form and Growth, Monograph N° 4, J.A. Mc Namara ed., Cranio-Facial Growth Series, Center for Human Growth and Development, Univ. Michigan, Ann Arbor, 275 p.

Stutzmann J (1976). Particularités de la croissance post-natale des cartilages secondaires du squelette facial. Recherches *in vivo* et en culture organotypique, chez le jeune rat, sur les processus de commande et de régulation. Thèse de Doctorat d'Etat ès-Sciences, Univ. Louis Pasteur, Strasbourg (France), 230 p.

Stutzmann J, Petrovic A (1974). Le muscle ptérygoïdien externe, un relais de l'action de la langue sur la croissance du condyle mandibulaire. Données expérimentales. Orthod Fr 45:385.

Stutzmann J, Petrovic A (1978). Einfluss von Testosteron auf die Wachstumsgeschwindigkeit des Kondylenknorpels der jungen Ratte. Rolle des "Vergleichers" des Servosystems welches die Verlängerung des Unterkiefers kontrolliert. Fortschr der Kieferorthop 39:345.

Grants C.R.L. INSERM N°76.1.135.4
       A.T.P. INSERM N°42.76.74.13

CONTROL OF RHYTHMS

THE ROLE OF SUPRACHIASMATIC NUCLEUS AFFERENTS IN THE CENTRAL REGULATION OF CIRCADIAN RHYTHMS

W.J. Rietveld and G.A. Groos

Department of Physiology and Physiological Physics, University of Leiden, Wassenaarseweg 62, 2300 RC Leiden, Netherlands

The suprachiasmatic nuclei of the mammalian hypothalamus (SCN) are considered to play a major role in the central control of circadian rhythms. Lesions of the SCN as well as transection of SCN efferents result in severe disruption of circadian rhythmicity (Rusak and Zucker, 1979). A major afferent projection to the SCN originates in the retina. This retinohypothalamic projection has been shown to mediate photic entrainment of circadian rhythms to the daily light-dark (LD) cycle (Rusak and Zucker, 1979, Groos and Mason, 1980). However, from these facts it can not be inferred that the SCN contain an independent endogenous pacemaker for the circadian system in mammals. To warrant such a conclusion, it is necessary to show that the SCN exert **their** neural control over the circadian system independently of their humoral environment and afferent neural inputs. Whereas it is difficult to control the humoral environment of the SCN, the contribution of SCN afferents to circadian timekeeping is relatively simple to evaluate experimentally. In addition to the retino-hypothalamic projection SCN afferents originate in the ventral part of the lateral geniculate nucleus (LGN_v; Hickey and Spear, 1976), the midbrain raphe nuclei (Aghajanian et al., 1969, Rusak and Zucker, 1979), the medial septal nucleus (MSN; Garris, 1979), the preoptic area (POA; Swanson, 1976) and the ventromedial (VMH) and arcuate nuclei of the hypothalamus (ARC; Makara and Hodács, 1975, Záborsky and Makara, 1979).

The effects of raphe-complex lesions on circadian rhythms are well documented. Although these nuclei send a prominent serotoninergic projection to the SCN such lesions do not prevent normal entrainment or freerunning of circadian

rhythms (Block and Zucker, 1976, Kam and Moberg, 1977). In
the present study we concentrated on the role of the remaining septal, hypothalamic and geniculate afferents of the
SCN.
In 20 Wistar rats bilateral electrolytic lesions were placed
in the VMH (cathodal current, 1.2 mA x 20 sec.), while in 10
rats such lesions were induced by chronic implantation of 30
µg goldthioglucose. Neurotoxic lesions of the arcuate nuclei
(ARC) were induced by neonatal subcutaneous administration
of monosodium-L-glutamate (2.0 up to 4.2 mg/g body weight
for 10 consecutive days) in 18 rats. The POA was lesioned
electrolytically in 9 rats (0.75 mA x 20 sec.). Septal afferents to the SCN were destroyed electrolytically by MSN
lesions (0.75 mA x 30 sec; n = 3) or interruption of the
projection at the level of the anterior hypothalamic area
(AHA, 0.75 mA x 25 sec; N = 4; Garris, 1979). Electrolytic
lesions were also made bilaterally in the LGN_v of 3 animals
(0.75 mA x 25 sec.) and the POA of 10 animals (0.75 mA x 15
sec.; n = 6 or 1.2 mA x 30 sec., n = 4). For all lesions an
adequate number of sham operated control rats were prepared.
In all cases stereotaxic surgery was performed under either
pentobarbitone sodium (60 mg/kg i.p.) or Hypnorm^R (1 ml/kg
i.m.) anaesthesia. Microscopic examination of the brains indicated that all lesions had been successfully placed. In the
case of VMH lesions the occurrence of the hyperphagia further
testified to the effectiveness of the lesions. Before and
after surgery the animals were housed individually in light,
temperature- and humidity-controlled rooms with free access
to food and water. Their food intake was recorded continuously by monitoring the number of food approaches in 30-
minute intervals (Rietveld et al., 1979). The standard
lighting conditions consisted of daily cycles of 12 hours of
darkness and 12 hours of light. In some experiments this regimen was phase advanced by 6 hours to study the rate of reentrainment to a newly imposed LD cycle. To test for the
endogenous nature of the circadian rhythms in some animals,
freerunning rhythms were induced by binocular enucleation.
Eating activity in unlesioned rats was predominantly nocturnal with 75-80% of the daily food intake consumed in the
dark when the animals were entrained to the standard LD
cycle. This normal circadian pattern and its phase angle
difference with the LD cycle is preserved after lesions of
the MSN, AHA, LGN and ARC (e.g. Fig. 1). In the case of VMH
lesioned rats, however, the concomitant hyperphagia somewhat obscured the normal feeding pattern. A change in the
amplitude of the rhythm was also observed in 4 rats sus-

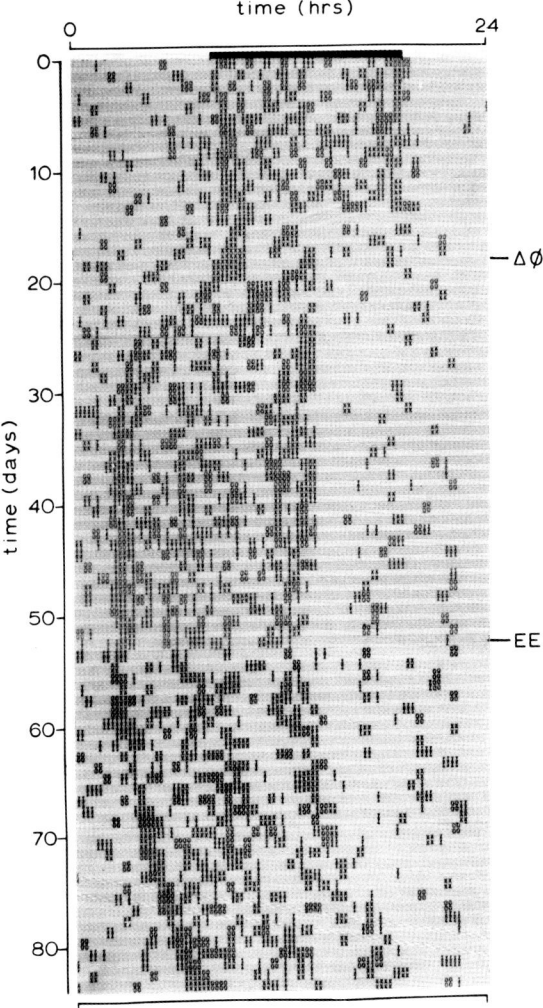

Fig. 1. Circadian food intake rhythm of an AHA lesioned rat. Consecutive daily records are plotted vertically. The dark bar above the record indicates the dark portion of the LD cycle before the lighting regimen was phase advanced by six hours ($\Delta\phi$). After binocular enucleation (EE), the animal freeruns.

taining large lesions of the POA. In these animals the nocturnality was more pronounced after the operation (Fig. 2). The basic properties of a circadian pacemaker are its intrinsic period and phase while the entrainment mechanism can be adequately characterized by the phase-response curve for short light pulses (Daan, 1977). During entrainment to LD cycles, the interaction between pacemaker period and phase-response curve results in a stable phase angle difference

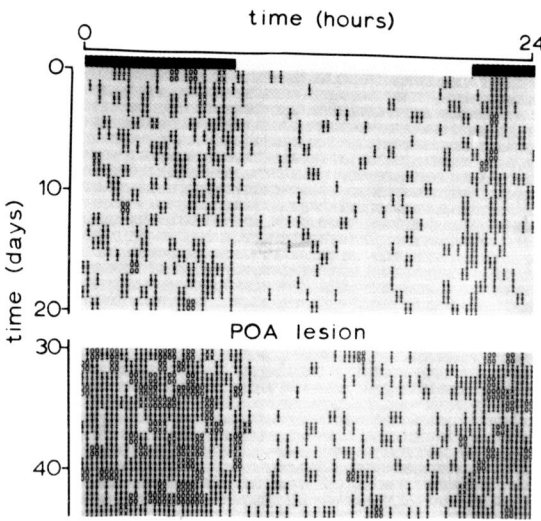

Fig. 2. Food intake rhythm of a rat before (days 0 - 20) and after a POA lesion (days 30-45). Note the marked increase in rhythm amplitude of the nocturnal eating activity.

between the LD cycle and the overt rhythm. From our observation that the phase relation of the food-intake rhythm to the LD cycle is not altered by lesions of SCN afferents, it can be inferred that the lesions did not affect the entrainment mechanism. This is particularly interesting in the case of the geniculate projection to the SCN which may be assumed to be visual (Rusak and Zucker, 1979, Groos, 1980) and therefore could contribute to photic entrainment. Our results would contradict such an assumption. The absence of an effect of SCN afferent lesions on entrainment was further illustrated by the re-entrainment experiments performed on MSN, AHA, POA and ARC lesioned rats. These animals normally re-entrained their rhythms within 5 days as did control animals (Figs. 1,3). In the same animals we observed freerunning rhythms after enucleation with periods exceeding 24 hours (mean and SEM: 24.26 $\pm$ 0.14 hrs in lesioned rats; 24.31 $\pm$ 0.17 hrs in control rats). The stability of the freerunning rhythms was normal as assessed by linear regression analysis of the onset times of the

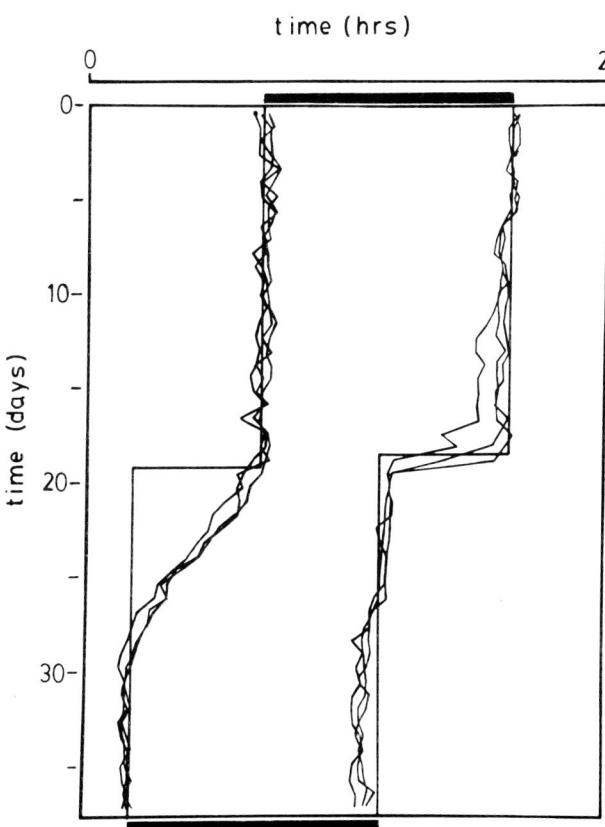

Fig. 3. Re-entrainment of the circadian food intake rhythm in POA, AHA and sham lesioned rats. In this record the averaged beginning and end of the nocturnal eating activity are plotted for each day. The dark bar above the record indicates the initial dark phase of the LD cycle. The new LD cycle is indicated at the bottom of the record.

nocturnal eating bout. Normal lengthening of the freerunning period was also observed in these lesioned animals after addition of 20% D_2O to the drinking water.
These findings, taken together with earlier observations obtained from rats sustaining lesions of the midbrain raphe complex, indicate that lesions of structures projecting to the SCN do not alter the basic properties of the pacemaker of the circadian food-intake rhythm. This pacemaker is assumed to reside in the SCN (Rusak and Zucker, 1979) which, although it receives various afferent neuronal inputs, apparently exerts its circadian control over the food intake of the rat independently of these projections. This conclusion is surprising as it raises the problem of what the function of these SCN afferents is if they do not participate in the control of circadian rhythmicity. Especially in the case of the LGN_V projection to the SCN, it is attractive to assume a role in the entrainment process. Further studies should be undertaken to establish whether these afferents modulate circadian rhythmicity in a more subtle way than by controlling phase or period. In the case of the larger POA lesions, evidence for such an effect was obtained. POA lesioned animals showed an increased L:D ratio of their food intake. At present it is difficult to decide whether this effect is due to amplitude control of the food-intake rhythm by the POA via its afferents to the SCN or to a direct influence on the homeostatic control mechanism of food intake in the mediobasal hypothalamus. Alternatively it is possible that the SCN afferents have no role in circadian timekeeping by the SCN but mediate in the neuroendocrine functions of these nuclei (Rusak and Zucker, 1979). Regardless of what the function of the SCN afferents may turn out to be, it is important that the control of the SCN over the circadian system is largely independent of its afferents. In this respect the present findings emphasize the status of the SCN as an endogenous circadian pacemaker.

REFERENCES

Aghajanan GK, Bloom FE, Sheard MH (1969). Electron microscopy of degeneration within the serotonin pathway of rat brain. Brain Res. 13: 266.
Block M, Zucker I (1976). Circadian rhythms of rat locomotor activity after lesions of the midbrain raphe nuclei. J. comp. Physiol. 109: 235.

Daan S (1977). Tonic and phasic effects of light in the entrainment of circadian rhythms. Ann. N.Y. Acad. Sci. 290: 51.

Garris DR (1979). Direct septo-hypothalamic projections in the rat. Neurosci. Lett. 13: 83.

Groos GA, Mason R (1980). The visual properties of rat and cat suprachiasmatic neurones. J. Comp. Physiol. 135: 349.

Groos GA (1980). An electrophysiological study of the suprachiasmatic nucleus. Chronobiologia, in press.

Hickey TL, Spear PD (1976). Retinogeniculate projections in hooded and albino rats: an autoradiographic study. Exp. Brain Res. 24: 523.

Kam LM, Moberg GP (1977). Effect of raphe lesions on the circadian pattern of wheel running in the rat. Physiol. Behav. 18: 213.

Makara GB, Hodács L (1975). Rostral projections from the hypothalamic arcuate nucleus. Brain Research 84: 23.

Rietveld WJ, ten Hoor F, Kooij M, Flory W (1979). The use of food hoppers for monitoring feeding habits of rats. Z. Versuchstierk. 21: 136.

Rusak B and Zucker I (1979). Neural regulation of circadian rhythms, Physiol. Rev. 59: 449.

Swanson LW (1976). An autoradiographic study of the efferent connections of the preoptic region in the rat. J. comp. Neurol. 169: 227.

Zabõrszky L, Makàra GB (1979). Intrahypothalamic connections: an electron microscopic study in the rat. Exp. Brain Res. 34: 201.

TIMING OF THE ESTROUS CYCLE IN RATS. ENDOGENOUS PEROXIDASE
ACTIVITY IN THE HYPOTHALAMIC ARCUATE NUCLEUS AS A TOOL IN
CIRCUIT ANALYSIS

W.J. Rietveld, E. Marani and J.C. Osselton

Dept. of Physiology and Psychological Psychology,
Dept. of Anatomy, State University of Leiden,
Wassenaarseweg 62, 2300 RC Leiden, Netherlands.

Horseradish peroxidase is employed in retrograde studies of
neuronal afferents to a certain area. Endogenous peroxidase
activity is assumed to be of minor importance because the
activity is easily abolished (Marani, 1980a). However this
is not true in the case of certain hypothalamic areas which
contain fixative resistant peroxidase activity (Rietveld,
1979a, **1979b; Marani, 1979). Forty-eight hours** immersion in
4% glutaraldehyde is required to destroy this activity.
After enzyme histochemical processing of formalin or glutaraldehyde perfused brains, **their reaction products are readily**
apparent (while other peroxidase activity is abolished by
this fixation treatment)(Marani, 1979).
A multidisciplinary approach was chosen to study the hypothalamic peroxidase activity, which was found to fluctuate.
Detailed descriptions and methodology are given in our
earlier reports (Rietveld, 1979a; Marani, 1978; Rietveld,
1979c). Several of the available techniques, including the
diamino benzidine and p-phenylene diamine methods, were
used in our study of   hypothalamic peroxidase. Light
and electron microscopy revealed the reaction product confined to clusters of granules (Marani, 1979; Marani, 1980a;
Rietveld 1979d) within hypothalamic areas. In mature rats
these clusters are found in the arcuate nucleus. They are
also present in the ependymal cells lining the ventricle.
The tetramethyl benzidine method, interestingly, differentiated between arcuate and ependymal cells by showing
reaction products only over the ependymal cells. The yield
of peroxidase reaction product under varying substrate concentrations, pH of the incubation media and different
glutaraldehyde fixations showed that this peroxidase ac-

tivity is probably of the catalase type (Marani, 1979). The presence of certain other enzymes in the catalase positive granules could point towards the involvement of a particular subcellular mechanism (Marani, et al. 1980c). Enzyme histochemistry showed that acid phosphatase, a lytic enzyme, cytochrome oxidase, end-enzyme of the respiration chain, and the neurotransmitter-converting enzymes monaminoxidase and acetylcholinesterase are not present. Lysosomal, mitochondrial or neurotransmitter conversion therefore seem improbable. However, glutamate dehydrogenase was demonstrated in granules within the arcuate nucleus and absent in control sections treated with inhibitors lacking the substrate or heated before staining (Marani, et al. 1980b). Studies at various postnatal ages (Rietveld, et al. 1979d) have demonstrated a displacement of the catalase reaction products from the median eminence (ME) via an intermediate area, into the arcuate nucleus (ACN). The catalase activity is first recognized in the ME on postnatal day 20-22 in normal rats and was exclusively present in the ACN at day 55. At day 37 catalase activity is found in the intermediate area and beginning to invade the ACN. Peroxidase histochemistry at the ultrastructural level of 20-day-old and mature rats showed in the ME and ACN respectively, catalase activity in 800-1000 $\overset{o}{A}$ granules, which are bounded by a single membrane. These granules were present in neuroblast-like or tanycyte-like cells (Marani et al. 1979; Rietveld, et al. 1979d). The cytoplasm around the karyon in both the young (catalase positive in ME) and adult (positive in ACN) rat contains these positive granules, and they can also be distinguished in cell protrusions (Fig. 1,A,B). On the typical neuroblast-like cells are found, in close proximity, dense-core boutons. These boutons are present at both ages, thus in both hypothalamic areas (Fig. A and inset of Fig. 1B).

In the ultrastructural study of young animals, ependymal cells bore no obvious dense-core boutons but were loaded with catalase positive granules (Fig. 1B). Mature rats showed these positive granules in cell protrusions near blood vessels, which is suggestive of some endocrine mechanism (Marani, et al. 1979).

Neonatal administration of monosodium glutamate (MSG) showed in mature rats destruction of the cells in the dorsolateral part of the arcuate nucleus (Rietveld, et al. 1979c). However, the catalase-containing cells in the ventro-medial part are unharmed. The catalase positive cells are believed to survive MSG treatment because of their content of a

Fig. 1A: ACN cell with catalase positive (open circles) 800-1000 nm granules surrounded by typical dense-core boutons (arrows). Inset: part of a cell protrusion loaded with catalase positive granules.

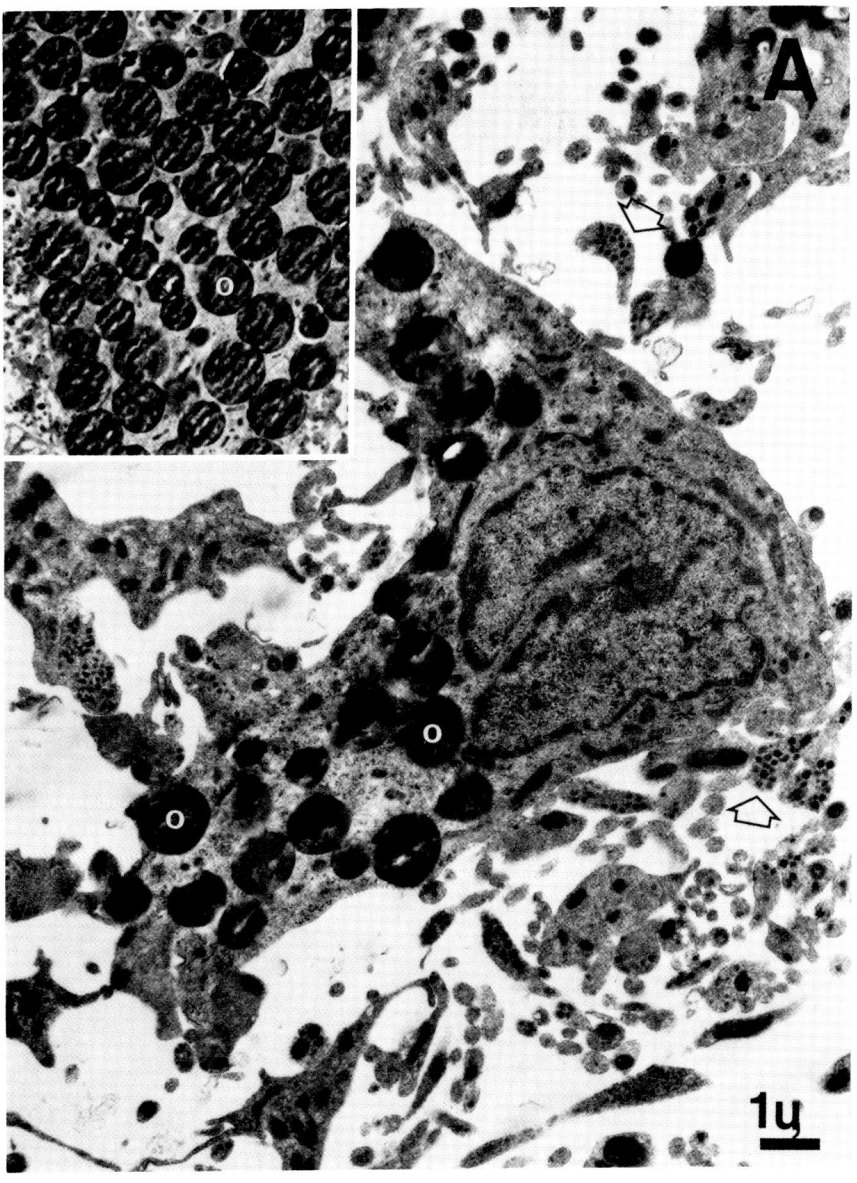

Fig. 1B: Ventricular lining cells with catalase positive 800-1000 nm granules (open circles). Note the absence of dense-core boutons. Inset: cell protrusion containing catalase positive granules (open circles) in the ME, directed from the ME towards the intermediate area. Dense-core boutons (arrows) surround this cell protrusion.

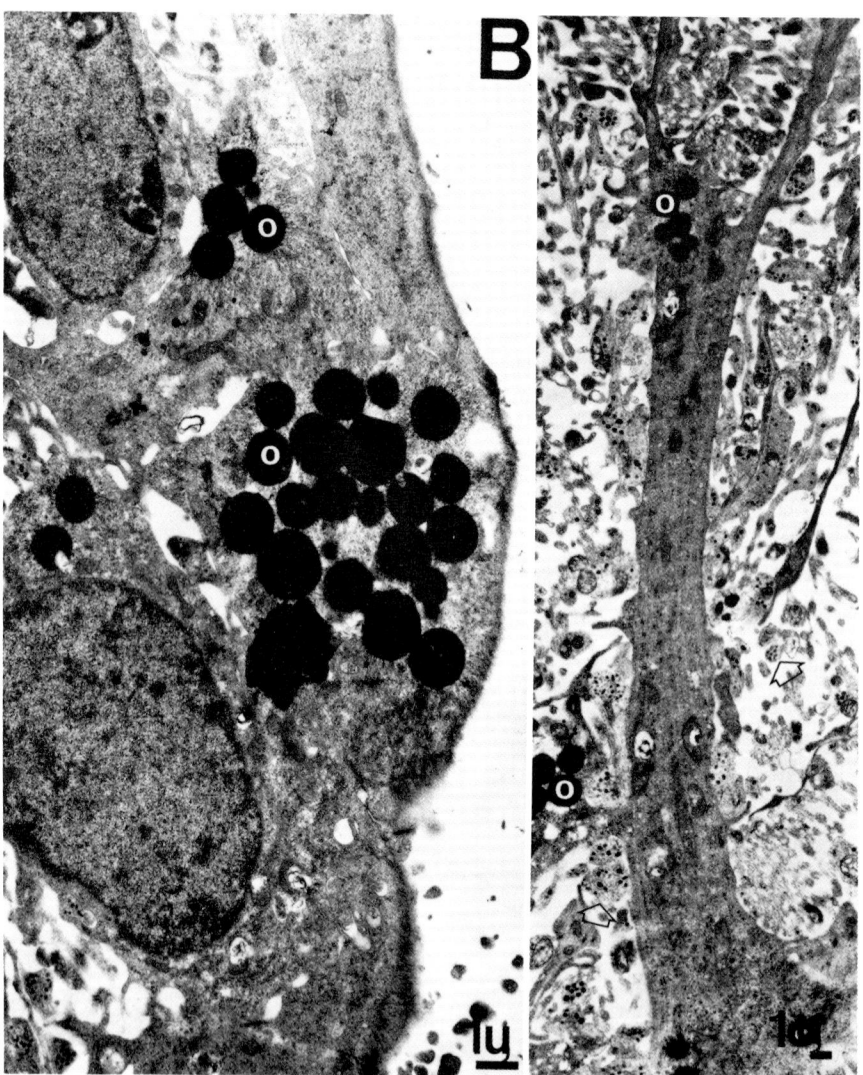

natural antidote, glutamate dehydrogenase (Marani, et al. 1980c). The MSG-treated rats are in persistent estrus and have a reduced body weight combined with an increased body fat content as compared to untreated animals (Rietveld, et al. 1979c). It is generally accepted that the anterior hypothalamic area is responsible for the phasic behaviour in LH-RF release (van Rees, 1972). The absence of this fluctuation in the presence of a normal LH-RF content in MSG treated animals could be explained if cells of the dorsolateral ACN (the area 12 in which dopaminergic cells are known to lie) mediate in this periodic release (Rietveld, et al. 1979c).

The regulation centers of circadian and estrous rhythms are located in the anterior hypothalamic area (Rietveld and Groos, 1980). The suprachiasmatic nucleus (SCN) seems to play a key role in this area (van Rees, 1972) and is considered to be a "pace-maker" for circadian rhythms. Moreover the SCN also influences the female cycle, because after sectioning of SCN connections the estrous cycle is disturbed (Rietveld and Groos, 1980). The catalase activity of mature male rats was measured at four different points in the 24 hrs cycle, and no indication could be found for a circadian rhythm (Rietveld, et al. 1979b). Changes that were perceived in catalase activity coincided with different stages in the estrous cycle. It therefore was concluded that the arcuate catalase activity in female mature rats was estrus-dependent (Rietveld, et al. 1979a). These results point to a relation between catalase activity and LH-RF. LH-RF too is found in the arcuate nucleus and in 800-1000 $\overset{o}{A}$ granules, is estrus related and catacholamine sensitive (Rietveld, et al. 1979b). Catalase activity thus is related to factors of the LH-RF system, whether to LH-RF itself or to a postulated intermediate protein (de Koning, 1976), and can be considered as a marker for the LH-RF metabolism in the ACN (Marani, et al. 1979). This is further supported by the coincidence of the appearance of catalase positive 1000 $\overset{o}{A}$ granules in the ACN with the onset of puberty in female rats, as measured by the opening of the vaginal orifice.

A hypothesis concerning the onset of puberty was proposed (Marani, et al. 1980d) using a simple neuronal circuit (Fig. 1C). It is based on the likelihood that **the SCN is the** pace-maker of the estrous cycle (Rietveld and Groos, 1980). The SCN can govern the ACN, and it is proposed that this action is exerted via the dorsolateral part of the ACN. The dopaminergic, MSG sensitive cells of this part of the ACN

Fig. 1C. Schematic representation of the circuitry studied. For explanation see text. Abbreviations: SCN: suprachiasmatic nucleus, ACN-DL: arcuate nucleus, dorsolateral part, ACN-VM: arcuate nucleus, ventromedial part, ME: median eminence.

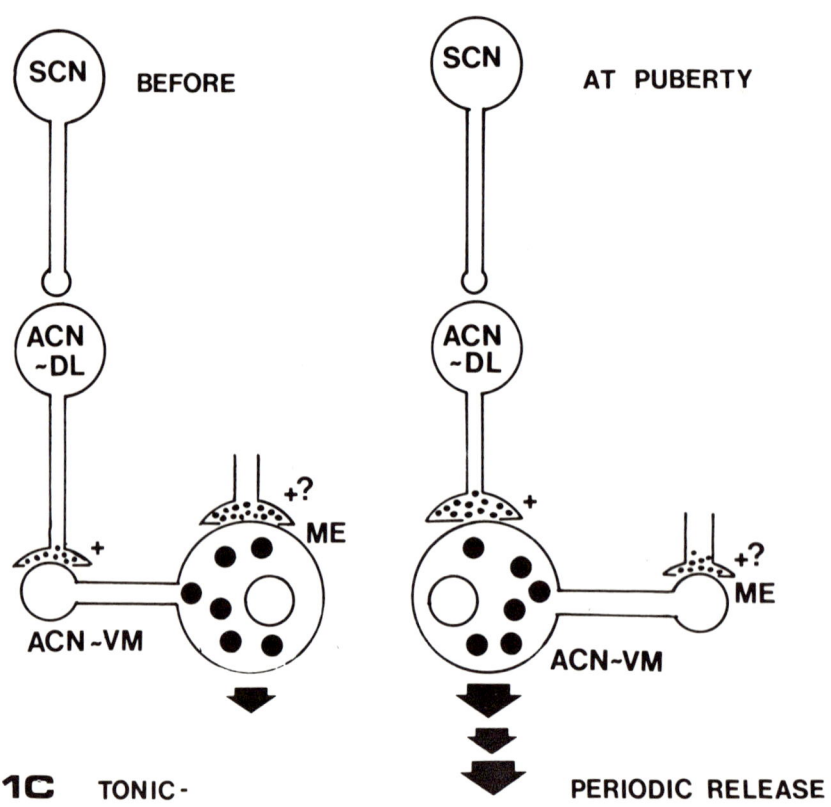

activate the ventromedial, catalase-containing cells and thus the related LH-RF system (Marani, et al. 1980d). Studies on catalase localization show that before puberty these catalase, LH-RF related, granules are outside the influence of the SCN, producing a tonic LH-rf release. At the onset of puberty, the displacement of the nucleus and perikaryal cytoplasm into the ACN makes it possible for the SCN to exert its influence, resulting in a periodic LH-RF release in young animals (33 days postnatal). Golgi studies of the ACN confirm the presence of bipolar cells with one protrusion in the ME and the other one in the ACN. Since the specificity of LH-RF antibodies is currently a matter of great concern (Sofroniew, et al. 1979; van Leeuwen, et al. 1978; and personal communication with these authors) any localization of LH-RF within the hypothalamus is questionable. All authors agree on the presence of LH-RF within the ACN, but there remains serious doubt as to the type of cells involved. In order to temporarily avoid the problem of specificity, our experiments were directed towards castration effects on catalase activity.

In the literature there abound confusing statements concerning identification of catalase activity and other histochemical products, like Gomori positivity (Marani, et al. 1980b). Gomori positivity is reported to change in castrated animals. However, a biomathematical analysis shows it to be very unlikely that Gomori positivity and catalase activity are identical. Effects on Gomori positivity are therefore of only marginal relevance to our study.

Chemical castration by administration of testosterone proprionate to neonatal female rats accelerates catalase displacement (Osselton, et al. 1980). In castrated animals the catalase activity is demonstrable in the ME at an earlier age and it arrives sooner in the ACN, indicating that probably a negative feed-back mechanism is released.

Since female rats were castrated, while females and males served as controls, a trigger mechanism via ovarian hormones was postulated. Studies are in progress on catalase changes in the arcuate nucleus of mature castrated animals (Rietveld, et al. 1980e). In conclusion the endogenous arcuate catalase activity seems to be related to the regulation of the rat's sexual hormonal activity both in onset and cyclicity.

Since the catalase activity is easily demonstrated, its study can be regarded as a useful tool in the investigation of puberty mechanisms. Perhaps even catalase activity denominates not only neuronal network changes but also intragranular mechanisms.

REFERENCES

De Koning J, van Dieten JAMJ, van Rees GP (1976). LH-RH dependent synthesis of protein necessary for LH release from rat pituitary glands in vitro. Molecular and Cellular Endocrinology 5: 151.

Van Leeuwen FW, Swaab DF, de Raay C (1978). Immunoelectronmicroscopic localization of vasopressin in the rat suprachiasmatic nucleus. Cell Tissue Res. 193: 1.

Marani E (1978). A method for three-dimensional reconstruction. Stain Technology 53: 265.

Marani E, Rietveld WJ, Osselton JC (1979). Ultrastructural localization of the endogenous peroxidase activity in the ventromedial arcuate nucleus. ICRS Medical Science 7: 501.

Marani E (1980a). Enzyme histochemistry. In Lahue R (ed): "Methods in Neurobiology": Plenum Co. p 183.

Marani E, Spoor CW, Rietveld WJ, den Hoed JL, Osselton JC (1980b). Is catalase positivity identical to the Gomori positivity in the arcuate nucleus? IRCS Medical Science, in press.

Marani E, Rietveld WJ, Mohanlal R, Osselton JC (1980c). Glutamate dehydrogenase is present in granula in the hypothalamic arcuate nucleus. IRCS Medical Science, in press.

Marani E, Rietveld WJ (1980d). The onset of puberty is characterized by the appearance of peroxidase activity in the arcuate nucleus. Anat. Anzeiger, in press.

Osselton JC, Rietveld WJ, Marani E (1980). Effect of castration on hypothalamic catalase displacement in prepuberal rats. IRCS Medical Science, in press.

Van Rees GP (1972). Control of ovulation by the pituitary gland. In: Ariens Kappers, Schade (eds): "Progress in Brain Research" Vol. 38: Elsevier Co, p 193.

Rietveld WJ, Osselton JC, Verwoerd N, Ploem JS (1979a). Endogenous peroxidase activity in the arcuate nucleus of the hypothalamus in rats. IRCS Medical Science 7: 176.

Rietveld WJ, Osselton JC, van Hylkema J, van Ingen E, Ploem JS (1979b). Endogenous peroxidase activity in the arcuate nucleus of the hypothalamus at different times of the day. IRCS Medical Science 7: 346.

Rietveld WJ, Osselton JC, Verwoerd N, van Ingen EM (1979c). The effect of monosodium glutamate on the endogenous peroxidase activity in the hypothalamic arcuate nucleus in rats. IRCS Medical Science 7: 573.

Rietveld WJ, Marani E, Osselton JC (1979d). Postnatal displacement of endogenous peroxidase activity from median eminence towards the ventromedial arcuate nucleus at puberty. IRCS Medical Science 7: 617.

Rietveld WJ, Marani E, Osselton JC (1980e). Effect of castration on hypothalamic catalase activity in mature rats. IRCS Medical Science. In press.

Rietveld WJ, Groos GA (1980). The central neuronal regulation of circadian rhythms. In: Scheving L, Halberg F (eds): "Principles and application of chronobiology to shifts in schedules with emphasis in man". Alphen a/d Rijn: Sijthof, in press.

Sofroniew MW, Weindl A, Schinko, Wetlstein R (1979). The distribution of vasopressin-, oxytocin-, neurophysin-producing neurons in the guinea pig brain. Cell Tissue Res 196: 367.

CHRONOBIOLOGICAL ASPECTS OF THE MAMMALIAN PINEAL GLAND

Russel J. Reiter

Department of Anatomy
The University of Texas Health Science Center
7703 Floyd Curl Drive
San Antonio, Texas 78284 U.S.A.

INTRODUCTION

A number of rhythms characterize the pineal gland of mammals (Reiter, 1975). Within the gland indole metabolism is highly cyclic with the maximal synthetic activity occurring during the dark phase of the light:dark cycle (Klein, 1979). The rhythms in indole metabolism appear to be circadian inasmuch as they persist in animals kept under continuous darkness (Wurtman et al., 1968). Besides the 24-hour rhythms there may be circannual rhythms in pineal activity as well (Mogler, 1958; Griffiths et al., 1979) especially in mammals exposed to natural photoperiodic and temperature conditions. The present survey will concentrate on the circadian rhythms in indole metabolism within the pineal of mammals, particularly rodents.

CYCLIC EVENTS WITHIN THE PINEAL GLAND

A great deal is known concerning indole metabolism within the pineal gland (Klein, 1979). The parent compound for the indoles is tryptophan. This amino acid is taken up from the blood by pinealocytes, and it is converted to serotonin and eventually to melatonin (Fig. 1), a major secretory product of the gland. Tryptophan is initially metabolized to 5-hydroxytryptophan by an enzyme which exhibits a high degree of activity within the pineal gland, namely, tryptophan hydroxylase (Lovenberg et al., 1968; Deguchi, 1972). The hydroxylated derivative is then

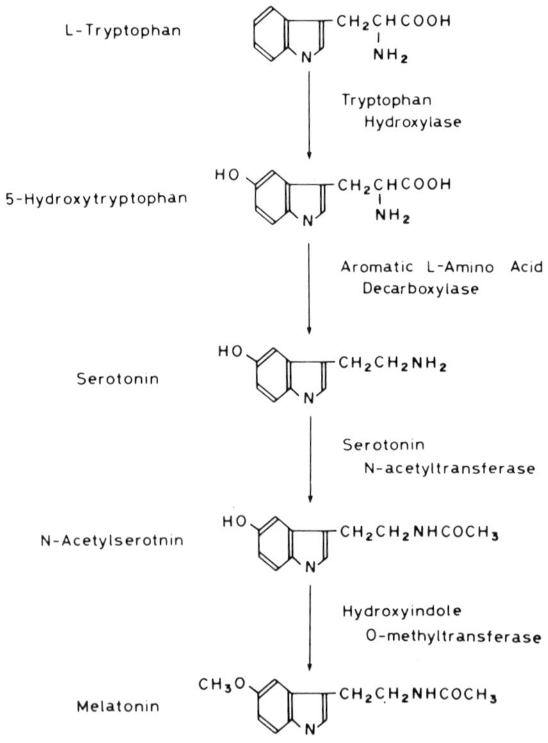

Fig. 1 Tryptophan metabolism within the pineal gland.

decarboxylated by aromatic amino acid decarboxylase with the resultant formation of 5-hydroxytryptamine (serotonin). The concentration of sertonin within the pineal gland exceeds that in any other organ in the body (Giarman, Day, 1959). There are also two steps involved in the conversion of serotonin to melatonin. First, N-acetylserotonin is formed following the N-acetylation of serotonin by the enzyme serotonin N-acetyltransferase (NAT) (Klein, Weller, 1970); the synthesis of N-acetyl-5-methoxytryptamine (melatonin) is then catalized by the enzyme hydroxyindole-O-methyltransferase (HIOMT) (Axelrod, Weissbach, 1960). The methyl group in this latter conversion is provided by S-adenosylmethionine. Besides melatonin a number of other indoles are formed during the metabolism of serotonin, e.g., 5-methoxytryptophol and 5-methoxytryptamine. Although these have received less interest as potential secretory products of the pineal gland, some of them may yet prove to be endocrinologically important (McIsaac et al., 1964; Vaughan et al., 1972).

The conversion of serotonin to melatonin is a highly cyclic event which is closely related to the prevailing light:dark cycle to which the animals are exposed. In all animals thus far studied, melatonin production is greatest within the pineal gland during the dark phase of the light:dark cycle (Quay, 1964; Lynch, 1971; Panke et al., 1978). During the period of light the level of serotonin is at its highest. With the onset of darkness the activity of NAT increases substantially (Klein, Weller, 1971; Rudeen et al., 1975); this results in the conversion of serotonin to N-acetylserotonin causing an increase in this constituent within the pineal (Klein, 1979). At the same time, because it is being metabolized, serotonin levels drop (Quay, 1963). Whether the activity of the melatonin forming enzyme also exhibits a rhythm currently is being debated (Klein, 1979). Regardless of whether it does or does not, maximal melatonin production within the pineal gland occurs during the dark phase of the photoperiodic cycle (Lynch, 1971; Panke et al., 1978; Tamarkin et al., 1979). The dependency of the increased melatonin production on the period of darkness is easily demonstrated. For example, if animals at middark are exposed to an artificial light source there is a rapid and profound drop in pineal melatonin levels (Tamarkin et al., 1979; Rollag et al., 1980). Under normal alternating periods of light and darkness, at the time of lights on in the morning, NAT activity and melatonin concentrations within the pineal gland have returned to near daytime levels. The daily rhythms in pineal serotonin, NAT, and melatonin in mammals are truly circadian inasmuch as they persist under conditions of continual darkness or after the animals are surgically blinded.

The increased production of melatonin within the pineal gland is mirrored by similar rises in plasma and urinary levels of this constituent. Indeed, there is a high degree of correlation between pineal and blood levels of melatonin when they are measured simultaneously

(Wilkinson et al., 1977). In all species in which these paramters have been examined, blood levels (Hedlund et al., 1977; Kennaway et al., 1977) and urinary excretion (Ozaki, Lynch, 1976) of the indole follow a precise circadian cycle with the period of darkness always being associated with peak melatonin levels.

Figure 2 shows the type of melatonin rhythm which occurs in the pineal gland of the Syrian hamster. Regardless of the photoperiod to which the animals are exposed, i.e., light:dark cycles of either 14:10, 12:12, 6:18 or 2:22, the period of darkness is accompanied by an accumulation of radioimmunoassayable melatonin within the pineal gland. If the light:dark cycle is reversed, after a brief period of adjustment, the melatonin rhythm is also reversed. If animals are exposed to continual light, the rise in melatonin production never occurs. From the data presented in figure 2, it is obvious that the length of darkness has little effect on the duration of peak melatonin production.

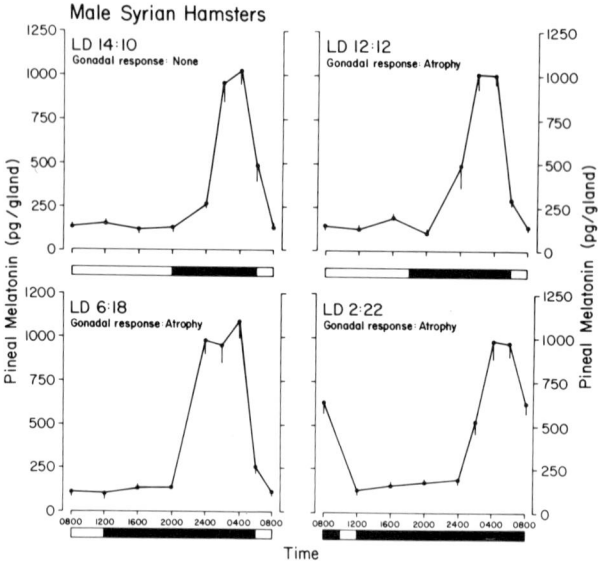

Fig. 2 Pineal melatonin rhythm throughout a 24-hour period in hamsters kept under light:dark (LD) cycles of either 14:10, 12:12, 6:18, or 2:22. Three of 4 light:dark cycles led to gonadal atrophy.

From the data presented in figure 2, it appears that the length of the daily dark period has little effect on the duration of elevated melatonin production. The failure of the length of the dark period to influence appreciably the amount of melatonin produced is of great interest to individuals working in the field, since hamsters in 3 (12:12, 6:18; 2:22) of the 4 photoperiods displayed experience gonadal involution if maintained under these conditions for a period of 8 to 10 weeks (Reiter, 1980). Conversely, the sexual organs of hamsters maintained under photoperiodic cycles of 14:10 remain reproductively competent. The gonadal involution which occurs in hamsters placed under short days is believed to be due to the production and secretion of melatonin by the pineal gland (Reiter, et al., 1978; Reiter, 1980). Since the amount of melatonin seemingly does not increase in animals kept under short days then there must be another explanation for the resulting reproductive collapse. Thus, it is presently assumed that peak melatonin production and secretion must coincide with the sensitivity of the animals to melatonin, i.e., there are 2 rhythms which must be in phase before darkness will induce gonadal atrophy in hamsters. It is already known that the reproductive organs of hamsters are only sensitive to exogenously administered melatonin for a restricted portion of each 24-hour period (Tamarkin et al., 1976; Reiter, 1980). Hence, the rhythm in melatonin production and the rhythm in the sensitivity of the melatonin receptors to the indole must be synchronized before darkness can induce reproductive involution. This synchrony is only achieved in hamsters kept under short day conditions. The interactions whereby darkness controls the reproductive status of hamsters may be applicable to other species as well.

FACTORS WHICH ALTER PINEAL MELATONIN RHYTHMS

Any procedure which interferes with the cyclic production of melatonin within the pineal gland also negates the ability of this organ to inhibit reproductive physiology. For example, continual light exposure prevents the daily rise in melatonin levels (Fig. 3) within the pineal and also renders the gland impotent in terms of the reproductive system. Likewise, destruction of the sympathetic nerve supply to the pineal by bilateral superior cervical ganglionectomy obliterates cyclic melatonin production (Panke et al., 1979) and prevents short days from inducing gonadal involution (Reiter, Hester, 1966).

Interruption of the postganglionic sympathetic nerves to the pineal obviously prevents the gland from receiving information about the photoperiodic environment. The system of fibers connecting the retinas to the pineal gland have been at least partially identified and are known to involve the sympathetic division of the autonomic nervous system (Moore, 1978). The neurotransmitter released from the post ganglionic sympathetic neurons at the level of the pinealocytes is known to be norepinephrine (Zatz, 1979). The neurotransmitter acts on β-adrenergic receptors to stimulate a series of biochemical events which culminate in melatonin production. Because of this, the administration of a β-aderenergic blocking agent (e.g., propranolol) prevents darkness from stimulating melatonin synthesis (Fig. 3).

Considering the profound influence which the pineal exerts over the reproductive system, it was anticipated that gonadal steroids would probably also have an inhibitory feedback effect on cyclic melatonin production. Thus, it was not altogether unexpected when Cardinali et al. (1976) reported the presence of estrogen and testosterone receptors in pineal tissue and claimed that the steroids influenced melatonin production in the rat. These findings were consistent with an earlier observation by Quay (1964), also using rats, that melatonin synthesis was lowest at the time of estrus. Because the pineal of the hamster is particularly active in terms of regulating gonadal function we deemed it important to determine the potential feedback influence of circulating ovarian steroids throughout the estrous cycle on the circadian rhythm of melatonin production. Since the levels of circulating steroids fluctuate markedly during various phases of the estrous cycle, we reasoned that melatonin levels would be changed accordingly. When subjected to test, however, neither daytime nor nighttime melatonin levels varied during the various stages of the estrous cycle (Fig. 4) (Rollag et al., 1979). Thus, unlike in rats, plasma steroids may not be involved in determining pineal indole metabolism.

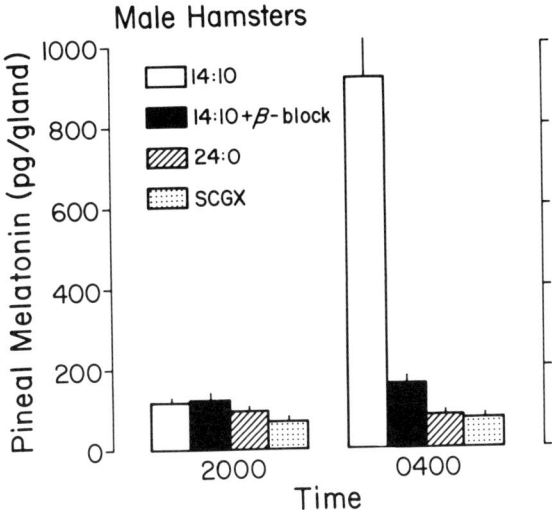

Fig. 3 Influence of light:dark cycles of 14:10 (with or without the administration of propranolol, a β-adrenergic blocking agent), constant light exposure (24:0), or superior cervical ganglionectomy (SCGX) on the daily melatonin rhythm in the pineal. Animals at 2000 h were killed during the light; those at 0400 h were killed during the dark.

SUMMARY

Pineal melatonin production is highly cyclic and depends on alternating periods of light and darkness. Peak melatonin synthesis occurs during the dark phase of the light:dark cycle. Interruption of the sympathetic nerve supply to the pineal gland, exposure of the animals to continuous light, or the administration of a β-adenergic antagonist obliterates the daily rise in pineal melatonin levels. At least in hamsters, circulating gonadal steroids seem to have minimal influence on the circadian rhythm in melatonin production.

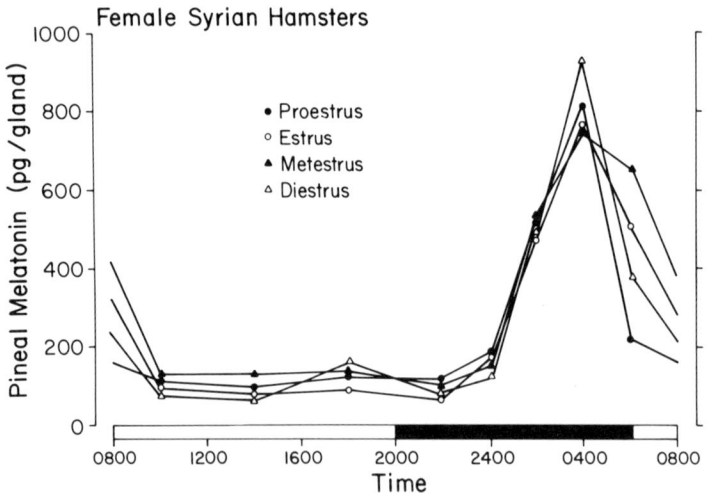

Fig. 4 Failure of the various stages of the estrous cycle to alter the 24-hour rhythm in pineal melatonin production. Standard errors are not included since they would have complicated the figure unduly.

REFERENCES

Axelrod J, Weissbach H (1960). Enzymatic O-methylation of N-acetylserotonin to melatonin. Science 131: 1312.

Cardinali DP, Nagle CA, Rosner JM (1976). Pineal-gonadal relationships: Nature of the feedback mechanism at the level of the pineal gland. In Anand Kumar TC (ed): "Neuroendocrine Regulation of Fertility," Basel: Karger, p. 206.

Deguchi T (1972). Tryptophan hydroxylase in pineal gland of rat: Postsynaptic localization and absence of circadian changes. J Neurochem 18: 667.

Giarman NJ, Day M (1959). Presence of biogenic amines in the bovine pineal gland. Biochem Pharmacol 1: 235.

Griffiths D, Seamark RF, Bryden MM (1979). Summer and winter cycles in plasma melatonin levels in the

elephant seal (Mirounga leonina). Austr J Biol Sci 32: 581.
Hedlund L, Lischko MM, Rollag MD, Niswender GD (1977). Melatonin: Daily cycles in plasma and cerebrospinal fluid of calves. Science 195: 686.
Kennaway DJ, Firth RG, Phillipou G, Matthews CD, Seamark RF (1977). A specific radioimmunoassay for melatonin in biological tissue and fluid and its validation by gas chromatography-mass spectrometry. Endocrinology 101, 119.
Klein DC (1979). Circadian rhythms in the pineal gland. In Krieger DT (ed): "Endocrine Rhythms," New York: Raven, p. 203.
Klein DC, Weller JL (1970). Indole metabolism in the pineal gland: A circadian rhythm in N-acetyltransferase. Science 169: 1093.
Lovenberg W, Jequier E, Sjoerdsma A (1968). Tryptophan hydroxylation: Measurements in pineal gland, brain stem, and carcinoid tumor. Science 155: 217.
Lynch HJ (1971). Diurnal oscillations in pineal melatonin content. Life Sci 10: 791.
McIsaac WM, Taborsky RG, Farrell G (1964). 5-Methoxytryptophol: Effect on estrus and ovarian weight. Science 145: 63.
Mogler RK-H (1958). Das endokrine System des syrischen Goldhamsters unter Berücksichtigung des natürlichen und experimentellen Winterschläf. Z Morph Oekol Tiere 47: 267.
Moore RY (1978). The innervation of the mammalian pineal gland. In Reiter RJ (ed): "The Pineal and Reproduction," Basel: Karger, p. 1.
Ozaki Y, Lynch HJ (1976). Presence of melatonin in plasma and urine of pinealectomized rats. Endocrinology 99: 641.
Panke ES, Reiter RJ, Rollag MD, Panke TW (1978). Pineal serotonin N-acetyltransferase activity and melatonin concentrations in prepubertal and adult Syrian hamsters exposed to short daily photoperiods. Endocr Res Commun 5, 311.
Panke ES, Rollag MD, Reiter RJ (1979). Pineal melatonin concentrations in the Syrian hamster. Endocrinology 104: 194.
Quay WB (1963). Circadian rhythm in rat pineal serotonin and its modifications by estrous cycles and photoperiod. Gen Comp Endocr 3: 473.

Quay WB (1964). Circadian and estrous rhythms in pineal melatonin and 5-hydroxyindole-3-acetic acid. Proc Soc Exp Biol Med 115: 710.

Reiter RJ (1975). Endocrine rhythms associated with pineal gland function. In Hedlund LW, Franz JM, Kenny AK (eds): "Biological Rhythms and Endocrine Function," New York: Plenum, p. 43.

Reiter RJ (1980). The pineal and its hormones in the control of reproduction in mammals. Endocr Reviews 1: 109.

Reiter RJ, Hester RJ (1966). Interrelationships of the pineal gland, the superior cervical ganglia and the photoperiod in the regulation of the endocrine systems of hamsters. Endocrinology 79: 1168.

Reiter RJ, Rollag MD, Panke ES, Banks AF (1978). Melatonin: Reproductive effects. J Neural Transmis, Suppl 13: 209.

Rollag MD, Chen HJ, Ferguson BN, Reiter RJ (1979). Pineal melatonin content throughout the hamster estrous cycle. Proc Soc Exp Biol Med 162: 211.

Rollag MD, Panke ES, Trakulrungsi WK, Trankulrungsi C, Reiter RJ (1980). Quantification of daily melatonin synthesis in the hamster pineal gland. Endocrinology 106: 231.

Rudeen PK, Reiter RJ, Vaughan MK (1975). Pineal serotonin-N-acetyltransferase in four mammalian species. Neurosci Letters 1: 225.

Tamarkin L, Westrom WK, Hamill AI, Goldman BD (1976). Effect of melatonin on the reproductive system of male and female Syrian hamsters: A diurnal rhythm in sensitivity to melatonin. Endocrinology 99: 1534.

Tamarkin L, Reppert SM, Klein DC (1979). Regulation of pineal melatonin in the Syrian hamster. Endocrinology 104: 385.

Vaughan MK, Reiter RJ, Vaughan GM, Bigelow L, Altschule MD (1972). Inhibition of compensatory ovarian hypertrophy in the mouse and vole: A comparison of Altschule's pineal extract, pineal indoles, vasopressin, and oxytocin. Gen Comp Endocr 18: 372.

Wilkinson M, Arendt J, Bradtke J, deZiegler D (1977). Determination of a dark-induced increase of pineal N-acetyltransferase and simultaneous radioimmunoassay of melatonin in pineal, serum and pituitary tissue of the male rat. J Endocr 72: 243.

Wurtman RJ, Axelrod J, Kelly DE (1968). "The Pineal Gland," New York: Academic.

Zatz M (1979). A neuropharmacological approach to the circadian oscillator regulating rat pineal serotonin N-acetyltransferase activity. In Suda M, Hayaishi O, Nakagawa H (eds): "Biological Rhythms and their Central Mechanism," Amsterdam: Elsevier, p. 149.

Index

Acid phophatase, 117-120, 128-129, 131
Acrophase, 14-15
Acrophase map, 44
  of mitosis, 49
  of S phase, 49
ACTH, 165, 177
Activity pattern, 8
Adjustment, occlusal, 196, 199-200
Adrenalectomy, effect on mitosis, 57
Adriamycin, 71
A-esterase, lysosomal, 27
Agar colony, 93
Alkylating agents, 143
Aminohydrolase, 88
Amitosis, 105
Amphetamine, 69
Amplitude, 2
Amylase in saliva, 12
Angle, Stutzmann's, 200
Antimycin-a, 32-33
Aphakia, 144
Aplasia, 145
Arabinosyl cytosine (ara-C), 66-67
Arcuate nucleus (ACN), 203-204, 214, 217-218
Area
  cell, 110, 112
  nuclear, 110, 113
Arylsulphatase, 117-119, 121, 123
Atrophy, gonadal, 226
Autonomic nervous system, 228

β-adrenergic blocking agent, 229
β-adrenergic receptors, 228
Basal temperature, 110
Bathyphase, 2
B-esterase, cytoplasmic, 27
β-glucuronidase, 26, 117-119, 122-123, 128-129, 130
Bile canaliculi, 119, 122-123
Binucleate cells, 29, 99, 101-102, 104-105
Biodisposability, 127
Blocking agent, β-adrenergic, 229
Blood cells, 87
Bone, endochondral trabeculae, 195, 197, 199
Bone marrow, 87, 94
  DNA in, 171, 173
  human, 43
Buccal cell, 110, 112
Buccal smears, 110-111

Cancer chemotherapy, 68
Cartilage, 188
  condylar, 193, 195, 198-200
  epiphyseal, 187-188, 190-191, 193
  primary, 187
  secondary, 187
Castration, 219
Catacholamine-sensitive cell, 217
Catalase
  activity, 217
  -containing cells, 219

-positive granules, 214–216
  reaction, 214
Cell area, 110, 112
Cell cycle, 81
  transit time, 84
Cell division, 30, 39, 50, 152
  in epidermis, 40
  in kidney, 40
  in pinna, 46
  in prepuce, 40
  in submandibular gland, 40
Cell proliferation in tumor, 167
Cell size, 112
Cells
  binucleate, 99, 101–102, 104–105
  blood, 87
  buccal epithelial, 109–110, 112
  catacholamine-sensitive, 217
  catalase-containing, 219
  dopaminergic, 217
  ependymal, 214
  erythropoietic, 87
  exfoliating, 109
  liver (see Hepatocyte)
  MSG-sensitive, 217
  myelopoietic stem, 94
  neuroblast-like, 214
  parabasal, 109
  precursor, 94
  squamous, 109, 111
  stem, 87, 94
  tanycyte-like, 214
  vaginal, 109–111
CFU_s, 87, 93
Chemotherapy, 167
  with adriamycin, 71
  with cyclophosphamide, 71
  schedules, 180
Chondroblasts, 188, 190–192
Chronobiology, definition, 1
  introduction to, 1
Chronogram, 2, 15
Chronotherapeutic treatment, 167
Chronotolerance to radiation, 152
Circadian pacemaker, 206, 208

Circadian rhythms, 3, 217
  control of, 203
  definition, 4
Circadian rule, 9
Circadian stage dependence in radiation, 151
Circannual rhythm, 3
Circaseptan rhythm 3, 177
Circatrigintan rhythm, 3
Cis-platinum, toxicity to, 72
Cleft palate, 144, 148
Colchicine, 45
Colony, agar, 93
Condylar cartilage, 187, 193, 195, 198–200
Condyle, mandibular, 195–200
Confidence arc, 15
Control of circadian rhythms, 203
Corneal epithelium, 47, 152–153, 155, 157, 159–160
Cornified cells, 109
Corpus luteum, 109
Corticosterone, 11
Cortisol, 11
Cosmic radiation, 5
Curve, phase response, 206
Cycle
  estrous, 213, 230, 338
  susceptibility-resistance, 158, 166–167, 177
Cyclophosphamide, 71, 143–144, 146–148
Cytometry, flow, 89
Cytoplasmic B-esterase, 27
Cytoplasmic size, 109
Cytosine arabinoside, 18
Cytostatic drugs, 34, 143

Decarboxylase, 224
Defects, ossification, 144
Densities, volume, 31
Dental arch, 196
Dimethyl benzanthracene, 61
Direction of growth, 195, 200
Disadaptation stress, 105

Diurnal rhythm, 4
DNA, 81, 110, 174
  bone marrow, 13, 67, 171, 173
  duodenal, 67
  esophageal, 13
  in Harding-Passey melanoma, 172
  hepatic, 104
  in peripheral blood, 13
DNA synthesis, effect of epidermal growth factor, 58–59
Dopaminergic cells, 217
Dose response, 155, 159
Drug susceptibility, 143, 146
Drugs, cytostatic, 34, 143

Embryotoxicity, 147–148
Endochondral bone trabeculae, 195, 197, 199
Endogenous peroxidase, 213
Endogenous rhythms, 5
Endoplasmic reticulum, 121, 123
  rough, 28
  smooth, 28, 33
Entraining agents, 5
Entrainment, 203, 206
Environmental factors, 190, 192–193
Enzyme activity, 118
Enzymes, hepatic, 99
Eosinophil, 53
  granulocytes, 88
Ependymal cells, 213–214
Epidermal growth factor, 58–62
Epidermis, 40
Epinephrine, 11
Epiphyseal cartilage, 187–188, 190–193
Epithelial cells, 109
Epithelium
  corneal, 152–153, 155, 157, 159–160
  squamous, 109
Error ellipse, 15
Erythrocytes, 87
Erythropoiesis, 87, 94
Erythropoietin, 88
Esterase, 127–129, 130, 134–136
Estrogen, 109, 228

Estrous cycle, 213, 217, 228, 230
Estrus, persistent, 217
Ethanol, 70
Exencephaly, 144
Exfoliating cells, 109

Fetal weight, 145
Flow cytometry, 89
Flow microfluorometry (FMF), 65, 81
Food-intake rhythm, 205, 207–208
Free-running, 6, 8, 16, 208
Frequency, 2
Frequency of labeled mitoses (FLM), 64–65
Fusion, nuclear, 105

G_1 phase, 83–84
G_2 block, 154
G_2 phase, 85
$G_2 + M$ phase, 83–84
Ganglionectomy, 228–229
Geophysical forces, 5
Gland
  pineal, 223
  salivary, 29, 31
  sublingual, 29
Glucagon
  effect on cell division, 61–62
  effect on DNA synthesis, 63
Glucosaccharo-1,4-lactone, 123
Glucose-6-phosphatase, 26, 28
Glutamate dehydrogenase, 214, 217
Glycogen, 26, 127
  hepatic, 99, 104
  phosphorylase, 26
  synthetase, 26
Golgi apparatus, 123
  lamellae, 121
Gonadal
  atrophy, 226
  involution, 227–228
  steroids, 228–229
Granules, catalase-positive, 214–216
Granulocytes, 87–88
Ground plasm, 119–120

Growth
  direction of, 195-196
  rate, 188-189, 192-193, 198-200
  rate, seasonal variations, 188
  rate, in tibia, 187
  retardation, 145

Harding-Passey melanoma, 165, 168-169, 174-176
  DNA of, 172
Hematopoietic tissues, 87
Hepatic
  DNA, 104
  enzymes, 99
  glycogen, 99, 104
  protein, 99, 104
  RNA, 104
Hepatocyte, 27, 31, 99, 117, 119-122
  binucleate, 29, 101
  mitotic activity, 105
  nuclei, 27
  volume, 27
Hexobarbital oxidase, 33
Histamine, 70
Histochemistry, lysosomal, 125
Hours of changing resistance, 152
Hours of changing susceptibility, 152
17-hydroxycorticosteroids, 16
Hydroxyindole-0-methyltransferase (HIOMT), 224
5-hydroxytryptamine, serotonin, 224
5-hydroxytryptophan, 223
Hydroxyurea, 165, 170, 175-176
Hyperphagia, 204
Hypophysectomy, effect on mitosis, 57
Hypoplasia, 145
Hypothalamic arcuate nucleus, 213
Hypothalamic peroxidase, 213
Hypothalamus, 203

Implantation sites, 144
Index, karyopycnotic, 110
Indole metabolism, 223
Indoles, 225
Indoxylacetate esterase, 26

Infradian rhythms, 3
Innate, 4
Insulin, effect on cell division, 61-62
Involution, gonadal, 227-228
Ionizing radiation, 151, 155, 160
Irradiation, 159
Isodose, 155
Isolation in a cave, 16

Jet-lag, 9

Karyopycnotic index, 110
Kidney, 40
Kinky tail, 144

Lateral geniculate nucleus, 203
LH-RF release, 217
LH-RF system, 219
Lidocaine, 70
Lighting duration, 193
Liver
  DNA, 29
  lysosomes, 128
  protein, 29
Longitudinal sampling, 3
L-tryptophan, 224
Lymphocytes, 87-88
Lymphocyte transformation, theory of, 41
Lysosomal enzymes, 27, 117, 129
Lysosomal esterase, 27, 135
Lysosomal histochemistry, 125
Lysosomes, 27, 117-119, 121, 123, 127

Magnetism, 5
Malformations, 143-148
Mandibular condyle, 195-200
Mandibular plane, 196
Marrow, bone, 87
Matings, timed, 144
Maturation index, 110
Maxilla, 195, 199
Meal timing, 33-34, 50, 52-54
Measurements, nuclear, 111
Mechanisms of rhythms, 56

Medial septal nucleus, 203
Median eminence (ME), 214, 218
Medullectomy, effect on mitosis, 57
Melanoma, Hardin-Passey, 165, 168–169, 174–176
Melatonin, 223–230
  receptors, 227
Menstrual rhythm, 111
Mesor, 14
Metabolism, indole, 223
5-methoxytryptamine, 225
5-methoxytryptophol, 225
Micrognathia, 145
Midbrain raphe nuclei, 203
Mitosis, 8, 151, 156, 160
Mitotic block, 160
Mitotic delay, 157, 159, 161
  radiation-induced, 154
Mitotic index, 152, 155–160
  ameloblast, 45
  cornea, 45, 51–52, 54
  duodenum, 45, 48
  effect of suprachiasmatic nucleus, 56, 58
  epidermis, 40, 45
  kidney, 40
  prepuce, 40
  submandibular gland, 40
  tongue, 45, 48
Mitotic rebound, 154
Monocytes, 87
Monosodium glutamate (MSG), 204, 214
MSG-sensitive cells, 217
Multiple-time-point sampling, 3
Myelopoiesis, 88, 94
Myelopoietic stem cells, 94

N-acetyl-t-methoxytryptamine (melatonin), 224
N-acetylserotonin, 224–225
NAT, 225
Nervous system autonomic, 228
Neuroblast-like cells, 214
Neurotransmitter, 228

Nicotine, 69
Nitrogen mustards, 143
Nocturnal, 4
Nonossification, 145
Norepinephrine, 228
Nuclear area, 110–112, 114
Nuclear-cytoplasmic area ratio, 110
Nuclear fusion, 105
Nuclear measurements, 111
Nuclear size, 105, 109, 112
$5'$-nucleotidase, 26
Nucleus
  arcuate (ACN), 203, 214, 217–218
  hypothalamic arcuate, 213
  lateral geniculate, 203
  medial septal, 203
  midbrain raphe, 203
  suprachiasmatic, 203, 217–218
  ventromedial, 203
Nyctohemeral variations, 187

Occlusal adjustment, 196, 199
Occlusion, 196
Ornithine decarboxylase, 12
Ossification defects, 144
Ovarian steroids, 109, 228
Ovary, 47
Ovulation, 109–110, 113

Pacemaker, 203, 208, 217
Parabasal cells, 109
Pentobarbital sodium, 18, 69
Period, 2
  freerunning, 208
Persistent estrus, 217
Phase, 2–3
  adjustment, 10
  -response curve, 206
  shift, 55
    by ara-C, 54
    of tumor, 167
Phases of cell cycle, 81, 83–84
Photic entrainment, 203
Pineal gland, 223
Pinealocytes, 223

Platelets, 87
Polar plot, 15
Post-irradiation recovery, 157
Postural hyperpropulsor, 196, 198
Precornified cells, 109
Precursor cells, 94
Precursor substances, uptake of, 105
Preoptic area, 203
Prepuce, 40
Primary cartilages, 187, 192
Primary lysosomes, 117
Progesterone, 109
Projection, serotoninergic, 203
Prokaryocytes, 4
Proliferation, bone marrow, 94
Properties of rhythms, 4
Propranolol, 228–229
Protein, hepatic, 99, 104
Puberty, 217, 219
Purpose of biological rhythmicity, 6
Purpose of circadian system, 13

Radiation
  chronotolerance to, 152
  circadian stage dependence in, 151
  -induced block, 157
  -induced mitotic delay, 154
  ionizing, 151, 155, 160
  response, 158
Radiotherapy, 72
Ratio
  nuclear-cytoplasmic area, 110
  toxic-therapeutic, 117
Receptors
  β-adrenergic, 228
  melatonin, 227
  testosterone, 228
Recovery, post-irradiation, 157
Reentrainment, 204
Regulation of circadian rhythms, 203
Resistance, hours of changing, 152
Resistance-susceptibility cycles, 177
Restricted feeding (see Meal timing)
Resynchronization, 16
Retardation, growth, 145

Reticulocyte, 88
Retinohypothalamic projection, 203
Rhythm
  induction, 174–175
  uncoupling, 105
Rhythms, 4
  animals, 39
  blind people, 10
  cell division, 39
  estrous, 217
  food intake, 205, 207–208
  plants, 39
  seasonal, 35
  tumor, 165
RNA
  hepatic, 104
  polymerases, 105
Routh endoplasmic reticulum, 28

S-adenosylmethionine, 224
Salivary gland, 29, 31
Sampling
  longitudinal, 3
  transverse, 3
Sarcoma, submandibular, 61
Schedules, chemotherapeutic, 180
Seasonal rhythms, 35
Seasonal variations, 187, 195, 198
  in growth rate, 188
Secondary cartilage, 187, 192
Secondary lysosomes, 117
Self-sustaining rhythms, 6
Serotonin, 223, 225
  5-hydroxytryptamine, 224
  N-acetyltransferase (NAT), 224
Serotoninergic projection, 203
Shifts, staggered phase, 110
Single-time-point sampling, 3
Size
  cytoplasmic, 109
  nuclear, 105, 109
Sleep duration, 33
Smears
  buccal, 110–111
  vaginal, 110

Smooth endoplasmic reticulum, 28, 33
Societal regimen, 5
S phase, 45, 47, 83–85
 duration of, 64
Squamous cells, 111
Squamous epithelium, 109
Staggered phase shifts, 100
Stem cells, 87, 94
Steroids
 gonadal, 228–229
 ovarian, 228
Stress, disadaptation, 105
Strychnine, 69
Stutzmann's angle, 200
Sublingual gland, 29
Submandibular gland, 40
Submandibular sarcoma, 61
Succinate-cy c. reductase, 32
Succinic dehydrogenase, 26, 127
Suprachiasmatic nucleus (SCN), 56, 203, 217–218
Susceptibility, hours of changing, 152
Susceptibility-resistance cycle, 18, 151, 158, 160, 166–167, 177
Susceptibility rhythm to drugs, 66, 68–70, 146
Sympathetic nerves, 228
Synchronization, 50, 52–53, 165
 tumor, 165, 167–169, 177, 180
Synchronizer, 5, 127

Tancyte-like cells, 214
Temperature, 7, 187
 environmental, 190
 independence, 6
 in blind people, 10
 variations, 193
Teratogenesis, 143, 146–148
Testosterone, 114
 receptors, 228
Th-R, 143–144, 146–148
Thymidine kinase, 105

Tibia, growth rate, 187
Time cues, 5
Time of ovulation, 109, 113
Timed matings, 144
Tissues, hematopoietic, 87
Tongue epithelium, 81
Toxic-therapeutic ratio, 177
Toxicity to displatinum, 72
Trabeculae, endochondral bone, 195, 197, 199
Transmeridianal displacement, 10
Transverse sampling, 3
Treatment
 chronotherapeutic, 167
 postural hyperpropulsion, 198
Tryptophan, 223
Tryptophan hydroxylase, 223–224
Tumor
 cell proliferation in, 167
 -cell synchronization, 165, 167–169, 177, 180
 phase shift of, 167
 rhythms, 165

Ubiquitous rhythms, 4
Ultracytochemistry, 117
Ultradian rhythms, 3
Uncoupling, rhythm, 105

Vaginal
 cells, 109–112
 smears, 110
Variations
 nyctohemeral, 187
 seasonal, 187, 195
Ventromedial nuclei, 203
Volume densities, 31

Weight, fetal, 145

X-rays, 151

Zeitgebers, 5

PROGRESS IN CLINICAL AND BIOLOGICAL RESEARCH

Series Editors
Nathan Back
George J. Brewer

Vincent P. Eijsvoogel
Robert Grover
Kurt Hirschhorn

Seymour S. Kety
Sidney Udenfriend
Jonathan W. Uhr

Vol 1: **Erythrocyte Structure and Function,** George J. Brewer, *Editor*
Vol 2: **Preventability of Perinatal Injury,** Karlis Adamsons and Howard A. Fox, *Editors*
Vol 3: **Infections of the Fetus and the Newborn Infant,** Saul Krugman and Anne A. Gershon, *Editors*
Vol 4: **Conflicts in Childhood Cancer: An Evaluation of Current Management,** Lucius F. Sinks and John O. Godden, *Editors*
Vol 5: **Trace Components of Plasma: Isolation and Clinical Significance,** G.A. Jamieson and T.J. Greenwalt, *Editors*
Vol 6: **Prostatic Disease,** H. Marberger, H. Haschek, H.K.A. Schirmer, J.A.C. Colston, and E. Witkin, *Editors*
Vol 7: **Blood Pressure, Edema and Proteinuria in Pregnancy,** Emanuel A. Friedman, *Editor*
Vol 8: **Cell Surface Receptors,** Garth L. Nicolson, Michael A. Raftery, Martin Rodbell, and C. Fred Fox, *Editors*
Vol 9: **Membranes and Neoplasia: New Approaches and Strategies,** Vincent T. Marchesi, *Editor*
Vol 10: **Diabetes and Other Endocrine Disorders During Pregnancy and in the Newborn,** Maria I. New and Robert H. Fiser, *Editors*
Vol 11: **Clinical Uses of Frozen-Thawed Red Blood Cells,** John A. Griep, *Editor*
Vol 12: **Breast Cancer,** Albert C.W. Montague, Geary L. Stonesifer, Jr., and Edward F. Lewison, *Editors*
Vol 13: **The Granulocyte: Function and Clinical Utilization,** Tibor J. Greenwalt and G.A. Jamieson, *Editors*
Vol 14: **Zinc Metabolism: Current Aspects in Health and Disease,** George J. Brewer and Ananda S. Prasad, *Editors*
Vol 15: **Cellular Neurobiology,** Zach Hall, Regis Kelly, and C. Fred Fox, *Editors*
Vol 16: **HLA and Malignancy,** Gerald P. Murphy, *Editor*
Vol 17: **Cell Shape and Surface Architecture,** Jean Paul Revel, Ulf Henning, and C. Fred Fox, *Editors*
Vol 18: **Tay-Sachs Disease: Screening and Prevention,** Michael M. Kaback, *Editor*
Vol 19: **Blood Substitutes and Plasma Expanders,** G.A. Jamieson and T.J. Greenwalt, *Editors*
Vol 20: **Erythrocyte Membranes: Recent Clinical and Experimental Advances,** Walter C. Kruckeberg, John W. Eaton, and George J. Brewer, *Editors*
Vol 21: **The Red Cell,** George J. Brewer, *Editor*
Vol 22: **Molecular Aspects of Membrane Transport,** Dale Oxender and C. Fred Fox, *Editors*
Vol 23: **Cell Surface Carbohydrates and Biological Recognition,** Vincent T. Marchesi, Victor Ginsburg, Phillips W. Robbins, and C. Fred Fox, *Editors*

Vol 24:	**Twin Research,** Proceedings of the Second International Congress on Twin Studies, Walter E. Nance, *Editor*
	Published in 3 Volumes:
	    Part A: Psychology and Methodology
	    Part B: Biology and Epidemiology
	    Part C: Clinical Studies

Vol 25:	**Recent Advances in Clinical Oncology,** Tapan A. Hazra and Michael C. Beachley, *Editors*
Vol 26:	**Origin and Natural History of Cell Lines,** Claudio Barigozzi, *Editor*
Vol 27:	**Membrane Mechanisms of Drugs of Abuse,** Charles W. Sharp and Leo G. Abood, *Editors*
Vol 28:	**The Blood Platelet in Transfusion Therapy,** Tibor J. Greenwalt and G.A. Jamieson, *Editors*
Vol 29:	**Biomedical Applications of the Horseshoe Crab (Limulidae),** Elias Cohen, *Editor-in-Chief*
Vol 30:	**Normal and Abnormal Red Cell Membranes,** Samuel E. Lux, Vincent T. Marchesi, and C. Fred Fox, *Editors*
Vol 31:	**Transmembrane Signaling,** Mark Bitensky, R. John Collier, Donald F. Steiner, and C. Fred Fox, *Editors*
Vol 32:	**Genetic Analysis of Common Diseases: Applications to Predictive Factors in Coronary Disease,** Charles F. Sing and Mark Skolnick, *Editors*
Vol 33:	**Prostate Cancer and Hormone Receptors,** Gerald P. Murphy and Avery A. Sandberg, *Editors*
Vol 34:	**The Management of Genetic Disorders,** Constantine J. Papadatos and Christos S. Bartsocas, *Editors*
Vol 35:	**Antibiotics and Hospitals,** Carlo Grassi and Giuseppe Ostino, *Editors*
Vol 36:	**Drug and Chemical Risks to the Fetus and Newborn,** Richard H. Schwarz and Sumner J. Yaffe, *Editors*
Vol 37:	**Models for Prostate Cancer,** Gerald P. Murphy, *Editor*
Vol 38:	**Ethics, Humanism, and Medicine,** Marc D. Basson, *Editor*
Vol 39:	**Neurochemistry and Clinical Neurology,** Leontino Battistin, George Hashim, and Abel Lajtha, *Editors*
Vol 40:	**Biological Recognition and Assembly,** David S. Eisenberg, James A. Lake, and C. Fred Fox, *Editors*
Vol 41:	**Tumor Cell Surfaces and Malignancy,** Richard O. Hynes and C. Fred Fox, *Editors*
Vol 42:	**Membranes, Receptors, and the Immune Response: 80 Years After Ehrlich's Side Chain Theory,** Edward P. Cohen and Heinz Köhler, *Editors*
Vol 43:	**Immunobiology of the Erythrocyte,** S. Gerald Sandler, Jacob Nusbacher, and Moses S. Schanfield, *Editors*
Vol 44:	**Perinatal Medicine Today,** Bruce K. Young, *Editor*
Vol 45:	**Mammalian Genetics and Cancer: The Jackson Laboratory Fiftieth Anniversary Symposium,** Elizabeth S. Russell, *Editor*
Vol 46:	**Etiology of Cleft Lip and Cleft Palate,** Michael Melnick, David Bixler, and Edward D. Shields, *Editors*
Vol 47:	**New Developments With Human and Veterinary Vaccines,** A. Mizrahi, I. Hertman, M.A. Klingberg, and A. Kohn, *Editors*
Vol 48:	**Cloning of Human Tumor Stem Cells,** Sydney E. Salmon, *Editor*
Vol 49:	**Myelin: Chemistry and Biology,** George A. Hashim, *Editor*